D1276331

NONSTANDARD ANALYSIS

Nonstandard glasses

NONSTANDARD ANALYSIS

Alain Robert

University of Neuchâtel
Switzerland

Translated and adapted by the author

A Wiley–Interscience Publication

JOHN WILEY & SONS

New York • Chichester • Brisbane • Toronto • Singapore

First published as *Analyse non standard*, © 1985, Presses polytechniques romandes, Lausanne, Switzerland.

Library of Congress Cataloging-in-Publication Data:

Robert, Alain.
Nonstandard analysis.

Translation of: L'analyse non-standard.
'A Wiley–Interscience publication.'
Bibliography: p.
Includes index.
1. Mathematical analysis, Nonstandard. I. Title.
II. Title: Nonstandard analysis.
QA299.82.R5813 1988 519.4 87-27925
ISBN 0 471 91703 6

British Library Cataloguing in Publication Data:

Robert, Alain
Nonstandard analysis.
1. Mathematical analysis, Nonstandard
I. Title
515 QA299.82

ISBN 0 471 91703 6

Typeset by Macmillan India Ltd., Bangalore 560 025
Printed and bound in Great Britain by Anchor Brendon Ltd, Tiptree, Essex.

CONTENTS

PREFACE

ORIGIN OF NSA

It was in 1966 that A. Robinson's book [R] on nonstandard analysis (henceforth abbreviated NSA) appeared. In it, a first rigorous foundation of the theory of infinitesimals was developed. In fact, A. Robinson had been using theorems in mathematical logic in the fifties to derive known mathematical results in a nonclassical way. His methods were based on the theory of models and in particular on the Lowenheim–Skolem theorem. For example, he introduced extensions $^*\mathbb{N}$, $^*\mathbb{R}$, . . . of the classical number systems by looking at nonstandard models of their respective theories. Infinitely small and large numbers were to be found in the enlargement $^*\mathbb{R}$ of $\mathbb{R}$, but did not exist in $\mathbb{R}$ itself. In this way, he was able to justify proofs using infinitesimals and that was not possible before his discovery. Finally, his article [2] with Bernstein certainly showed that these methods were able to produce original solutions to unsolved mathematical questions as well.

Since 1977, the mathematical community has been able to approach this theory of infinitesimals without having to go through any tedious preliminary discussion in logic: the article [10] by E. Nelson shows indeed how logical foundations can be circumscribed in a neat axiomatic way. Thus, at least in principle, any mathematician can delve into NSA without prerequisites in formal logic (first order predicate calculus does not even have to be mentioned). To bring infinitesimals to life in the classical number systems, set theory itself is modified. Enriched with a few new deduction rules, 'special' elements can be found in all infinite sets. No enlargement is needed in this perspective. A new property of sets and elements is introduced, that of being 'standard' ('special' simply meaning 'nonstandard'). Suitably codified, this property reveals the existence of infinitesimals (and infinitely large numbers) in the classical set $\mathbb{R}$ of real numbers for example.

How can one give an exposition of NSA today without a preliminary chapter in logic? The answer to this question is simple enough: it is possible to the same extent as it is possible to develop classical mathematics, based on set theory, without starting with a chapter on axiomatic set theory. . . . Most mathematical texts start with an informal discussion about sets without even codifying any coherent system of axioms about them.

ZFC AND IST

Set theory was created by G. Cantor at the end of the last century. As many mathematicians noticed in the early twentieth century, antinomies could easily be produced by a careless use of the notion of set. A typical example is furnished by the 'catalogue of all catalogues not mentioning themselves': should one include this catalogue in itself?. . . . Both decisions lead to a contradiction!

There are several ways of avoiding paradoxes in set theory and several sets of axioms have been proposed to get rid of the famous antinomies of the above type. One of them is based on a distinction of 'types'. In an expression $x \in E$, the letter x denotes an element, whereas, the capital E denotes a set: these have different types, and expressions like $x \in x$, or $E \in E$ are not allowed. Another formalization of set theory is based on a restriction on relations that are set forming. Bourbaki gives a precise description of 'collectivizing relations', i.e. relations which can be used to define a set: the relation $x = x$ is not a collectivizing relation. Finally, in still another system, the Zermelo–Fraenkel axiomatic (henceforth abbreviated ZF), all relations are set forming in a pregiven set. Starting from a set E and a property P, there is always a set F (a part of E) consisting precisely of elements $x \in E$ having the property P. This is the specification axiom of ZF. In this theory, all objects under consideration are sets and expressions $x \in y$, $y \in z$, . . . are legitimate, but the relation $x = x$ can only be used to specify a set within a given set E.

E. Nelson introduces an 'internal set theory' based on ZF but uses one more term, namely the term 'standard'. We shall follow Nelson's point of view, based on ZF, but shall not give an extensive list of axioms of this theory (there are infinitely many since this theory has some 'axiom schemes'), only referring to them when needed. Let us simply recall that set theory is based on the binary relation $\in$. This symbol is not defined in set theory, but its use is codified by the axioms. However, a proper (naïve, or intuitive) interpretation for it is helpful! Similarly, the new term 'standard' used in internal NSA is undefined. Only the use of this 'predicate' is codified by the axioms. Nevertheless, a suitable interpretation of this term is useful. The main point is that it can be applied to any object of set theory (or simply to any set, since all objects of ZF are sets). Any set E is either standard or nonstandard, any object x in a set E is either standard or nonstandard, . . . any function is either standard or nonstandard, etc. This is the principle of the excluded middle. But of course, it may be difficult to discover which is true in specific examples.

HOW MUCH AXIOMATICS?

Since I claim that one can use NSA just as one uses set theory without starting with its axiomatics (who knows which axiomatics he is using in his everyday mathematical life?), I have to explain why I still have chosen an axiomatic way of presenting NSA. My purpose is not to formalize NSA. But one has to admit that since infinitesimals have been banished from sound mathematics for centuries, it would not be enough simply to state that they exist (or existed all along . . .). Those who have grown up in a mathematical world having no recognition of such entities will certainly experience an initial psychological difficulty: it would simply not be fair to content oneself with assertions that they exist. . . . On the other hand, it is good to realize that nonstandard elements are revealed in all infinite sets. This notion is thus not solely applicable to the numerical sets. Another reason for our choice is the following. This book is certainly not intended for a student unfamiliar with traditional mathematics and intuitive set theory. In this context, insisting on the change of perspective is a must, and this—I feel—is best done in a more dogmatic way. Changes of perspective must be founded on firm ground least they give an impression of arbitrariness or unsafety. In this book, NSA will be presented through the three deduction rules

(I) Idealization,
(S) Standardization,
(T) Transfer,

as first formulated by Nelson. To fix ideas, let us say that we work with the ZFC theory for sets (i.e. ZF with Choice axiom) to which we simply add the IST axioms. In particular, this means that we are not removing any of the results based on the ZF theory: all traditional mathematics based on set theory retain their validity in the new extended theory. To these classical statements, results, theorems, . . . we add those based on the new term 'standard' (its negation or any expression making an implicit use of it). This is the new internal set theory (IST for short).

However, it is quite possible that with a new mathematical generation, a naïve and intuitive approach to NSA might be possible or even better. After all, so many illustrious mathematicians—Leibniz, Euler and Cauchy in particular—all seem to have considered this approach more natural than the (nowadays called) traditional one.

COHERENCE

Playing the new game for a reader accustomed to the traditional deduction rules inferred from set theory will certainly create some uneasiness . . . . To what extent, for example, is it possible to introduce new axioms even if they concern a new term? Could they enter into conflict with the old ones? More precisely, are we assured that the new theory ZFC and IST is coherent, i.e. not contradictory? An interested reader could go back to the article [10] by Nelson where the following metatheorem (due to C. Hertz) is proved

> A classical statement which can be established with ZFC and IST can also be proved using the axioms of ZFC only.

In more technical terms, the extension of ZFC obtained by adjoining the IST axioms is conservative. Should we then, dismayed and disappointed, reject these new axioms as unproductive? Certainly not. A first reason for keeping them might be the following: a classical statement may have a shorter proof within the extended theory. As J. Hadamard had already observed concerning the use of complex numbers

> 'Sometimes, the shortest path linking two real facts goes through the complex domain!

A second reason is that the new theory allows for new statements. These would have no meaning in traditional mathematics but can still be interesting. Modelling reality on set theory only is certainly not a God-given rule to respect. On the contrary, the history of mathematics has shown that the intuition of great mathematicians could not be encompassed by set theory.

The above consistency statement has the following comforting meaning. Any mathematical contradiction is logically equivalent to the shocking

$$1=0$$

equality. This typical contradiction is a classical statement. If *a* contradiction could be proved in ZFC and IST, *this* classical contradiction could also be proved in ZFC and IST and hence, also in ZFC. In other words.

> If ZFC and IST is a contradictory theory, then ZFC is also a contradictory theory.

Although consistency of ZF is not proved, we work confidently within this theory. The same confidence can be attributed to ZFC and IST. Instead of insisting on these consistency results, our goal is to develop a feeling of confidence in the enlarged theory. Apparent difficulties over infinitesimals are resolved, and the reader should acquire the intuitive knowledge of how to do so.

INFINITESIMALS IN PHYSICS

Physicists will probably have the feeling that Robinson's or Nelson's way to infinitesimals leads away from their need. They like to think of infinitesimals in terms of differentials dx, df, . . . bearing no resemblance to nonstandard objects. One can be satisfied by the construction of the tangent variety, blowing up points into vector spaces. Tangent vectors do give an interpretation of infinitely close elements of the point. Jet varieties also give several orders of magnitude in infinitesimals. However, these constructions are only superficially different from the NSA approach which has wider applicability. For example, the notion of differential can only be applied to differentiable functions. The definition of differentiability itself cannot be based on a notion of differential dx without creating a vicious circle. Worse even: the notion of continuity cannot be discussed in the context of differentials dx! Another type of objection is the following: dx, the square of dx, etc. have a meaning. But what is the meaning of the dx-power of dx? Already Euler needed a whole algebra of infinitesimals (cf. Introduction).

FOR WHOM IS THIS BOOK WRITTEN?

We have already alluded to possible readers of this text. Let us come back in more detail to this point. This book is first aimed at advanced undergraduate students who have already practised a serious calculus course. Preferably, they should also have had some exposure to point set topology. It is probably when dealing with this topic that students first realize the meaning of abstract set theory: continuity is not a notion solely applicable to functions defined on numerical sets. These prerequisites already make it clear that students of this book must have a certain experience of mathematical abstraction. Logical preliminaries are kept to a minimum and it has been our guiding principle to avoid excessive formalization in dealing with the axiomatics. These limited preliminaries should be sufficient to approach the first part (Chapters 1 to 6) of this book. Perhaps one should remember that infinitesimals have never been abandoned in technical schools . . . . Mathematics teachers may often have had some hesitation over them, and some initial bad conscience. They should now learn that their practice can be reconciled with theory. The first part of this book is also intended for them.

Practising teachers, mathematicians and physicists should also be able to learn NSA from scratch here. For them, we have given a few applications in a second part of the book (Chapters 7 to 11). To a large extent, these chapters are independent from one another and can be read in any order. If a graduate student has never heard of p-adic numbers, he can skip the section mentioning them without being penalized later on. He can simply read the chapter on differential equations if that is his field of interest, or the chapter on invariant subspaces for polynomially compact operators. In each of these chapters of the second part, we have tried to emphasize one typical application of NSA.

EXERCISES

To acquire some confidence in working with NSA, it is necessary to solve some exercises (this is no particularity of this theory!). The change of perspective makes it even more imperative to check that the basic reflexes are properly assimilated. This is why I have included several exercises in the first crucial chapters, supplying some hints on their solutions (which should be read only if necessary) and finally giving complete answers. In this way, a beginner can really check his proper understanding of NSA: warnings and words of caution only become effective when practising! In later chapters, fewer details are given but the more difficult exercises giving for example NSA formulations of classical results are nevertheless treated in detail.

RESEARCH

Since more and more research is being made on NSA or using NSA, it was natural to write a didactic presentation of Nelson's point of view. We thus hope that it will attract the recognition that it deserves. Although our aim is primarily directed at communicating the basic ideas of NSA, a couple of examples and exercises are original and have not appeared in research periodicals.

We certainly do not claim completeness in any sense and have not been able to present all applications of the theory. To keep the book to a limited size, we have had to omit the following topics

—applications to probablity theory,
—stability in differential equations with small parameters (birth of 'ducks' according to the Strasbourg school),
—finite elements methods,
—integral points on curves (Siegel–Mahler theorem),
— . . . .

Perhaps the main applications of NSA are still to be found . . . . For example, the general principle asserting that any infinite set contains nonstandard elements could still furnish some interesting finiteness results. Even if these were obtained in a nonconstructive way, they might be interesting because of their simplicity.

BENEFITS

I shall not try to defend NSA for the simplifications that it can bring, nor for its aesthetic qualities: the reader can do it for himself! But let me simply observe that the novelty of the point of view can refresh a teacher's routine. So often we have taught that the set of rational numbers is countable, while the set of reals is not, so often we have repeated that the set of rationals is negligible in the set of reals (i.e. can be covered by a family of intervals, with an arbitrarily small sum of lengths) that it may be difficult for us to remember our initial doubts, difficulties and insecurities in handling these notions . . . . NSA provides us a new opportunity of experiencing insecurities, trivial mistakes . . . and thus reminds us that precise definitions and clear statements are insufficient for teaching a new field. Only practice and repetition can establish a firm foundation on which to build. It is only when one starts modifying set theory that one realizes the occult role that it plays. Perhaps a few misunderstandings can be avoided when meeting students' questions about real numbers.

EVOLUTION, NOT REVOLUTION

This text was initiated by a couple of courses that I had the opportunity to give to

—secondary school teachers (Les Paccots, Switzerland, 1981),

—Queen's Univ. (Kingston, Ont., Canada, 1982),
—analysis seminar (Univ. Neuchâtel, Switzerland, 1982–83 and 1986–87).

It was used by advanced undergraduate students, and it is typically intended for them. This text is not really intended for high-school teaching, but though it is not my place to offer advice—having never taught at that level—I would like to make the following observation: NSA can and should indirectly influence teaching methods, for example, in the handling of real numbers intuitively. Interested teachers will themselves be able to discover how much of NSA can be used with any benefit. Some interesting results have indeed already been obtained, but I have the impression that the experimentation period is not yet over.

Ultimately, NSA should be able to be used implicitely as freely as set theory has been: the language of set theory has filtered into all mathematics, but few mathematicians are aware of its axioms, much less of the different axiomatics for it. The same should eventually apply for NSA.

ACKNOWLEDGEMENTS

First, I have to thank the audiences who catalysed my presentation of this subject. They forced me to find examples and exercises in quantity, thus helping me to develop and convey a better feeling for the difference between standard and nonstandard. More specifically, I would like to thank E. Gilliard who carefully read a first manuscript of this book and detected several errors. My thanks also go to P. Cartier who thoroughly went through the French edition and who gave me a very useful list of misprints, suggestions, references, etc. Finally, I have one more debt towards my wife Ann, who checked my English.

Neuchâtel, June 1987 ALAIN ROBERT

CONVENTIONS, NOTATIONS

This book is divided into chapters. The first six chapters constitute a FIRST PART explaining the basic notions of nonstandard analysis. A SECOND PART, Chapters 7 to 11, presents a few applications of the theory. The chapters of this second part can be read independently and are based on a broader mathematical experience (roughly speaking, this second part is written for a postgraduate audience).

Each chapter is subdivided into sections by means of a second number. Most sections are subdivided into paragraphs. Thus;

> 4.5 refers to the fifth section of Chapter 4 (this section deals with 'Theorems on continuous functions')
> 4.5.3 refers to the third paragraph of the preceding section (it happens to consist of 'Comments' on the theorems).

Thus, a single linear numbering has been adopted, paragraphs representing alternatively results, proofs, comments, etc. A proof is always finished at the end of a paragraph (or at the end of a section if it has several parts!). Thus it was unnecessary to use a special symbol indicating the end of a proof. Occasionally, I use the abbreviation 'iff' for 'if and only if': it is ugly but irresistible . . . .

Technical terms are printed in bold characters when they are defined. They are listed in the index. Other terms appearing in italics should catch the attention of the reader: they have their usual intuitive meaning and are not listed in the index.

A few abbreviations are used, namely

- ZF: Zermelo–Fraenkel set theory,
- ZFC: Zermelo–Fraenkel set theory with axiom of choice,
- NSA: Nonstandard analysis,
- IST: Internal set theory (Nelson's point of view of NSA),
- I: Idealization axiom,
- S: Standardization axiom,
- T: Transfer axiom.

Bibliographical references are given by the initial of the author for books, and by a number for articles. The complete list of references—together with more information on the subject—is given at the end of the volume.

The fundamental numerical sets are denoted as follows

$\mathbb{N}=\{0, 1, 2, 3, \ldots\}$: set of natural integers,
$\mathbb{Z}=\{\ldots, -1, 0, 1, 2, \ldots\}$ additive group of rational integers,
$\mathbb{Q}=\{n/m : n\in\mathbb{Z}, 0\neq m\in\mathbb{N}\}$ field of rational numbers,
$\mathbb{R}=$ field of real numbers,
$\mathbb{C}=\mathbb{R}(\sqrt{-1})$: field of complex numbers.
$\mathbb{Z}_p=$ ring of p-adic integers $\subset \mathbb{Q}_p=$ field of p-adic numbers (when p is a prime).

Intervals of $\mathbb{N}$ or $\mathbb{R}$ are represented resp. by

$$[a, b] : a\leq x\leq b,$$
$$[a, b[: a\leq x< b, \text{etc.}$$

The inclusion symbol $A\subset B$ means '$x\in A \Rightarrow x\in B$' (thus our $\subset$ corresponds to the $\subseteq$ symbol also frequently used in America).

Maps between sets are simply denoted by arrows as in $f: E\rightarrow F$, the correspondence at the level of elements being indicated by a special arrow: $x\mapsto f(x)$.

FIRST PART

Ce qu'on voulait faire, c'est en le faisant qu'on le découvre. (Alain)

[*You find out what you really wanted to do when you are doing it*]

INTRODUCTION

Since L. Euler was among the most inspired users of infinitesimals, let him have the first word . . . .

Here is how he deduces the expansion of the cosine function in his book [E] (Caput VII, Section V 133.). He starts from the Moivre formula

$$\begin{aligned}\cos nz &= \tfrac{1}{2}[(\cos z + i\sin z)^n + (\cos z - i\sin z)^n]\\ &= \cos^n z - \frac{n(n-1)}{1\cdot 2}\cos^{n-2} z\sin^2 z\\ &\quad + \frac{n(n-1)(n-2)(n-3)}{1\cdot 2\cdot 3\cdot 4}\cos^{n-4} z\sin^4 z + \ \ldots\end{aligned}$$

and writes (*loc. cit.*, p. 141)

> *sit arcus z infinite parvus; erit* $\cos\cdot z = 1$, $\sin\cdot z = z$; *sit autem n numerus infinite magnus, ut sit arcus nz finitae magnitudinis, puta* $nz = v$
>
> $$\cos\cdot v = 1 - v^2/2! + v^4/4! - \text{etc.}$$

On the other hand, he had given explicit rules for dealing with *infinite* numbers (*loc. cit.*, p. 124)

$$\frac{i-1}{i} = 1, \qquad \frac{2i-1}{3i} = \frac{2}{3}, \ldots .$$

He also writes

$$\begin{aligned}e^x &= (1 + x/i)^i \quad i \text{ numerus infinite magnus}\\ &= 1 + i\frac{x}{i} + \frac{i(i-1)}{1\cdot 2}\frac{x^2}{i^2} + \ \ldots\\ &= 1 + x + x^2/2! + \ \ldots .\end{aligned}$$

Later (*loc. cit.*, pp. 125–126) he considers the *infinitesimal* $\varepsilon = 1/i$ to derive the series expansion of the logarithm

$$\begin{aligned} y = \ln x &\Leftrightarrow x = e^y = (1 + y/i)^i \\ &\Leftrightarrow x^{1/i} - 1 = y/i \\ &\Leftrightarrow y = \frac{1}{\varepsilon}(x^\varepsilon - 1); \end{aligned}$$

replacing x by $1+t$, we obtain the usual expansion:

$$\begin{aligned} y = \ln(1+t) &= \frac{1}{\varepsilon}[(1+t)^\varepsilon - 1] \\ &= \frac{1}{\varepsilon}\left[1 + \varepsilon t + \frac{\varepsilon(\varepsilon - 1)}{1 \cdot 2} t^2 + \ldots - 1\right] \\ &= t - t^2/2 + t^3/3 - \ldots . \end{aligned}$$

Of course, it is difficult to support equality of two numbers 1 and $(i-1)/i$ which are *different* even if i is infinitely large . . . .

Nonstandard analysis allows one to distinguish these two numbers and gives a precise meaning to

$$\frac{i-1}{i} \simeq 1 \quad (\textit{infinitely close} \text{ numbers}).$$

In particular, the notion of *infinitely close* numbers is *theoretically* well defined. It is also possible to define a notion of standard part (or *observable* part) for which

$$\operatorname{st}\frac{i-1}{i} = 1 \quad \text{if } i \text{ is infinitely large.}$$

(Since 1777, Euler reserves the letter i for the square root of -1 i.e. for the imaginary unit, and we shall follow him!)

When Euler did these 'formal' computations he certainly had a didactical motivation and one should imitate him on several occasions. For example, to introduce the number $e = 2.71828\ldots$, one can start from the computation of the derivative of the functions $x \mapsto a^x$. Consider a differential quotient relative to an increment $h \neq 0$:

$$\frac{a^{x+h} - a^x}{h} = a^x \cdot \frac{a^h - 1}{h}.$$

How should we choose the number a in order to have simple formulas? We can try to choose a so that

$$\frac{a^h - 1}{h} \simeq 1 \quad \text{when } h = 1/n \quad \text{and } n \text{ is infinitely large.}$$

In other words, we would like to have

$$n \cdot a^{1/n} - n \simeq 1,$$
$$a^{1/n} - 1 \approx 1/n,$$
$$a^{1/n} \approx 1 + 1/n,$$
$$a \simeq (1 + 1/n)^n.$$

Define thus

$$e = \mathrm{st}\,(1 + 1/n)^n = \lim\,(1 + 1/n)^n.$$

The symbols $\simeq$, $\approx$ are well defined in nonstandard analysis and the reader will find more examples of formal manipulations in this text. Fortunately, nonstandard analysis goes much beyond these —*a posteriori*—justifications!

Le charme: une manière de s'-entendre répondre 'oui' sans avoir posé de question claire!
(Albert Camus)

[*Charm: a certain manner of hearing the answer 'yes' when no precise question was asked*]

CHAPTER 1

IDEALIZATION

1.1 SET THEORY, NEW PREDICATE

1.1.1 (ZF) axiomatic of set theory

We shall adhere to the point of view that all traditional mathematics can be based on the **Zermelo–Fraenkel axiomatic** (henceforth abbreviated (ZF)) of set theory. Indeed, we know from experience that this formalization is well suited to our mathematical purpose .... This choice does not mean that we intend to advertize one—rather than another—axiomatic of set theory. But we have to fix ideas since a fundamental change is about to be made. However, our use of (ZF) will be intuitive (as opposed to formal) since our goal is 'mathematics'.

First of all, let us recall that all objects considered by (ZF) are 'sets'. For example, the number 0 is (or represents) the empty set $\varnothing$. Similarly, the number 1 is a set with one element: one can take the set $\{\varnothing\}$ having as only element the empty set that has just been mentioned. Set theory also uses a binary relation $\in$. The use of this relation is codified by rules which are precisely the axioms of set theory. Instead of memorizing these axioms, it is certainly better to have an intuitive interpretation of this symbol $\in$. One possible interpretation is that of 'belonging'. Thus if x and y are two sets, the relation $x \in y$ is interpreted as 'x belongs to y' (or 'x is an element of y'). This assertion $x \in y$ can of course be true or false, depending on the sets x and y! When we say that sets have 'elements', we are only adopting a terminology suited to the **interpretation** of $\in$ given above. In fact, all mathematical entities considered in set theory 'are' sets and no fundamental distinction should be made between the type of objects that can be placed either on the left, or on the right of the symbol $\in$. When we say 'element' instead of 'set', we only want to avoid terminological monotony .... It is true that in other set theories, distinctions of type between element and sets have been made, and we choose (ZF) to simplify these considerations!

As a typical example, let us mention the **axiom of extension**. This axiom states

$$\forall z(z \in x \Leftrightarrow z \in y) \Leftrightarrow x = y$$

which can be read

for any z (z belongs to $x \Leftrightarrow z$ belongs to y) iff $x = y$

(the usual abbreviation 'iff' stands for 'if and only if'). More concisely

two sets x and y are equal iff they have the same elements z.

Let us not review all axioms of set theory (since (ZF) is also based on some 'axiom schemes'; it has, in fact, infinitely many axioms . . .). Rather, we shall just mention them when needed, at some crucial points. In any case, the goal of these axioms is to construct mathematical sets. Equivalently, the axioms furnish a mathematical existence of usual sets.

As basic example, the set $\mathbb{N} = \{0, 1, 2, 3, \ldots\}$ of natural numbers consists of the positive integers

$$0 \in \mathbb{N}, \qquad 1 \in \mathbb{N}, \qquad 2 \in \mathbb{N}, \ldots.$$

The existence (and construction) of this set $\mathbb{N}$ is furnished by the axioms (ZF). This set is well and unambiguously defined within set theory and its meaning will be the same in Non Standard Analysis (henceforth abbreviated NSA).

1.1.2 New term 'STANDARD'

We shall adopt Nelson's point of view for NSA. It is based on all (ZF) axioms and some other axioms (or axiom schemes) concerning a new term. This new term **STANDARD** is a 'unary relation'—we shall say 'predicate'—which is part of the theory and can be applied to any set, i.e. to any mathematical object. Thus if x is a set, the assertion 'x is standard' is meaningful. It is a well-formed expression within NSA: as such, it can be either true or false. Since this new predicate is part of the new theory, it is not defined by any construction within (ZF). But its use has to be governed by certain rules or axioms, to be added to (ZF). The binary relation $\in$ was not defined either by (ZF): its use only is codified by the axioms of set theory. This is an essential point in any axiomatic: its objects are introduced before definitions. For example, in a topological space, the open sets are not defined, they are part of the given structure of topological space. The question 'is this subset open or not' always has a meaning in a topological space. Likewise, the question 'is this set standard or not' is always significant in NSA. And since all mathematical objects can be considered to be sets of (ZF), it is always legitimate to raise the question 'is this mathematical entity standard or not'.

Eventually, the use of axioms for decisions about 'standard or not' will be made obsolete by its intuitive interpretation. But since this intepretation is quite subtle—only due to the fact that its level of generality is so wide—we shall have to develop it gradually.

Functionally—or grammatically—the term standard should be compared to 'finite'. Indeed, 'finite' is also a qualificative which can be applied to any set. If x is a set, the question 'is x finite' is meaningful. It is a well-formed sentence and the answer can be either yes or no! The decision about which is true can however be quite difficult! The attributes 'finite' and 'standard' are not hereditary in the following sense. If $x \in y$ and y is finite, we cannot infer that x is finite. For example, the set $y = \{\mathbb{N}, \mathbb{Q}\}$ is finite since it has only two elements, but $x = \mathbb{N}$ is infinite although $x \in y$! Likewise, if y is a standard set and $x \in y$, the element x can very well be nonstandard. In other words, standard sets can contain nonstandard elements . . . and they do in most cases!

But in spite of the grammatical analogy, the attribute 'finite' is defined within set theory, whereas the attribute 'standard' is part of the language of NSA. The only way of proving that a specific set is standard (or not) is to have recourse to the axioms. These constitute the rules of the new game.

1.1.3 New versus old theory

One of the main advantages of Nelson's point of view—which we are adopting—is to retain all axioms of set theory. Consequently, all known mathematical theorems remain valid as such. All definitions keep their original meaning, all usual abbreviations are still in force, etc. Existing mathematics are not altered—they will be called **classical mathematics**—since they are based on (ZF) which will be embedded in NSA. New axioms will simply be *added* to (ZF) to govern the use of the new term 'standard'.

For example, the set $\mathbb{N}$ of natural numbers is the same in NSA simply because it is unambiguously defined in (ZF), and thus is part of our new system. More explicitly, we are *not going to add elements to the classical set* $\mathbb{N}$ *of natural numbers*, and we shall never refer to an 'extension $^*\mathbb{N}$' of $\mathbb{N}$ as Robinson initially did. But if $\mathbb{N}$ still represents the same classical set, it is also true that the new deduction principles—resulting from the new axioms—may give a psychological feeling of extension since they reveal elements that were unknown to (ZF). In a sense, the new axioms bring to life unsuspected elements in the traditional set $\mathbb{N}$. While this set $\mathbb{N}$ has not changed, people working with NSA discern 'more elements' in it, because they have a richer axiomatic. Of course, these unsuspected elements had always been there . . . . Finally, let us state firmly that if $n \in \mathbb{N}$ is any natural integer, either n is standard or n is nonstandard (excluded middle) and we must be capable of deciding, in each particular case, which is true. The axioms will impose that both types of integers indeed occur.

1.2 AXIOMATIC: BEGINNING

1.2.1 Abbreviations

Before giving the axioms (I), (S), (T) of NSA, let us introduce a few convenient abbreviations concerning quantifiers. Instead of the very formal

$$\forall x(x \in E \text{ and } x \text{ standard}) \Rightarrow \ldots$$

we shall simply write

$$\forall^S x \in E \ldots$$

and this should be read

for every standard element x of the set E, (we have) . . . .

Similarly, the following abbreviations

$$\forall^{sf} x \in E \ldots, \qquad \exists^s x \in E \ldots, \qquad \exists^{sf} x \in E \ldots$$

stand respectively for

for every standard and finite x of E (we have) . . .

there exists a standard element x in E such that . . .

there exists a standard and finite x in E such that . . .

When there is no possible confusion about the reference set E in which all elements are considered, we may even write more simply

$$\forall x \ldots \quad \text{instead of} \quad \forall x \in E \ldots \quad \text{or of} \quad \forall x, x \in E \Rightarrow \ldots$$

We do this since our goal is *mathematics* as opposed to *formal logic.* All quantifiers have their usual intuitive meaning. In fact, we shall progressively abandon the use of quantifiers altogether to approach a naïve use of NSA, thus imitating the usual naïve use of set theory. However, to avoid a psychological feeling of insecurity, it is perhaps best to rely on a precise formalism at the beginning.

1.2.2 Definition

Any statement, definition, formula, result, . . . which does not use (either explicitly or implicitly) the new predicate 'standard' will be called classical. Thus, 'classical' refers to sentences which can be formed in (ZF) (or traditional set theory).

1.2.3 Heuristic approach to idealization

The only purpose of this section is to give an intuitive approach to the first axiom of NSA. As such, its goal is not normative. But it should help in obtaining a proper understanding of idealization.

The nonstandard integers $x \in \mathbb{N}$ will correspond to the 'infinitely large' integers considered by Euler. These integers x should satisfy

$$x > 1, \qquad x > 1000, \qquad x > 10^{80}, \ldots$$

i.e. $x > n$ if n is a 'usual integer'. Usual integers would be the standard integers and we should postulate the existence of an $x \in \mathbb{N}$ with

$$x > n \quad \text{for all standard } n \in \mathbb{N}.$$

Such an x would, in particular, be different from all standard integers n, hence be nonstandard. Thus this postulate would imply the existence of nonstandard integers. In fact, it is better to give an axiom with wider applicability. For this we have to discover the essential property of the binary relation $R(x, n) =$ '$x > n$' permitting to require

$$\exists x \quad \text{with} \quad R(x, n) \quad \text{for all standard } n.$$

Nelson discovered that the basic property of the inequality that can be axiomatized is the following. To say that $x > n$ means that

$$x > y \quad \text{for all } y \in [0, n],$$

and if n is standard, all (finite) subsets of $[0, n]$ should also be standard. Typically, one should observe that for any finite and standard subset F of $\mathbb{N}$, say $F \subset [0, n]$, there is an

$$x \in \mathbb{N} \text{ with } R(x, y) \quad \text{for all } y \in F$$

(e.g. take $x = n + 1$). This property can be formulated for any classical binary relation of set theory (in place of $<$ in $\mathbb{N}$).

Physicists have long been using conditions like $x \gg 1$ for large numbers. Their notion is relative to the context (e.g. a life period of 10^{-10}s for the Δ particle is very large with respect to a 10^{-23}s mean life for an unstable Δ particle!). It could often be replaced by the corresponding theoretical notion from NSA.

1.2.4 Idealization axiom (I)

Let $R = R(x, y)$ be a classical binary relation. We postulate that the following two properties are equivalent

(i) for every standard and finite set F there is an $x = x_F$ such that $R(x, y)$ holds for all $y \in F$,
(ii) there is an x such that $R(x, y)$ holds for all standard y.

More formally this axiom is valid for classical binary relation $R(x, y)$ and states

$$\forall^{sf} F, \exists x = x_F [R(x, y) \forall y \in F] \Leftrightarrow \exists x [R(x, y) \forall^s y].$$

1.3 COMMENTS

1.3.1 Reformulation

To get an even better feeling for idealization, let us introduce the following notation. Instead of saying

$$R(x, y) \quad \text{holds for all } y \in F,$$

we shall simply write

$$R(x, F) \quad \text{holds.}$$

With this convention, the (I) axiom states that the classical relations R suited to idealization are precisely those for which

> for each standard and finite set F, there exists an x (depending on F) such that $R(x, F)$.

This property mimics majoration (or domination) and is equivalent by (I) to

> there is an x with $R(x, y)$ for all standard y.

1.3.2 Comments on binary relations

Mathematicians can imagine without any loss (!?) that the binary relations R to which idealization applies are defined with reference to sets E and E'. Such a relation is thus represented by (or even identified to) its graph—still denoted R—in the Cartesian product $E \times E'$. Thus, instead of saying that $R(x, y)$ is true, or holds, one can write $(x, y) \in R$. However, this is not strictly necessary and one could take for a classical binary relation the '$y \in x$' relation. In this particular case, the convention just introduced in (1.3.1) for the extension $R(x, F)$ represents the relation $F \subset x$. This classical relation has the characteristic domination property: if F is a finite standard set, we can take $x = F$ to satisfy $R(x, F)$ (obviously, this choice of x depends on F!). Consequently, (I) furnishes a set x for which any standard set y belongs to x.

We shall have to come back to this application and its interpretation (cf. (1.4.2) below).

1.3.3 A technical comment

The main interest of the idealization axiom (I) comes from the implication

> *if* R is a classical binary relation with the typical majoration property, *then*, there is an x such that $R(x, y)$ $\forall^s y$.

The converse implication is only needed for a technical point (cf. (2.4.3): in a standard finite set, all elements are standard).

1.4 FIRST APPLICATIONS OF (I)

1.4.1

Let us start with the classical binary relation $R(x, y)$ defined by

x and y are natural integers such that $x > y$.

In this case, $R(x, F)$ holds precisely when

the integer x is a strict majorant of the part $F \subset \mathbb{N}$.

When the part F is finite—*a fortiori* when it is standard and finite—it is possible to find an integer x with $x > y\ \forall y \in F$. Since F is finite, it is contained in an interval $[0, n]$ and we can take $x = n + 1$ (this choice indeed depends on F). Idealization (I) can be applied and asserts

there is an integer $x \in \mathbb{N}$ such that $x > y$ for all standard integers $y \in \mathbb{N}$.

In particular, as was already noticed in (1.2.3), this implies that nonstandard integers exist!

1.4.2

As a second application of (I), let us consider the classical relation $R(A, y)$

A is a finite part (of a set E) containing the element y.

Thus, $R(A, F)$ stands for

A and F are two parts (of E) with $A \supset F$.

If F is a (standard) finite part of E, we can take A (finite indeed!) equal to F so that $R(A, F)$ holds. Thus (I) is applicable and furnishes

there exists a finite part A of E containing all standard elements y of E.

Observe that (I) does not mean that there is a finite part A of E containing exactly the standard elements (all of them, no other) of E. In general, such a finite part will contain many nonstandard elements. If the predicate standard is interpreted by accessibility, the conclusion

A is finite and contains all accessible elements (together with some extraneous nonstandard elements)

certainly corresponds to a practical finiteness feeling for the notion of accessibility.

1.4.3

Finally, let us consider the relation $R = R(x, y)$ given by '$x \neq y$' in a certain set E. If $F \subset E$ the extended relation $R(x, F)$ simply stands for '$x \notin F$'. Let us assume that E is an infinite set so that for each finite subset F of E, we can find an $x \in E$ with $x \notin F$. In this case, (I) is applicable and furnishes an x with $x \neq y$ for all standard $y \in E$. In other words, x is nonstandard. We reach the conclusion that *in every infinite set E, there is a nonstandard element*. This is *one of the most important principles of NSA*.

In fact, if x is a nonstandard element of an infinite set E, the set $E' = E - \{x\}$ is still infinite and thus contains nonstandard elements. Thus, an infinite set E contains many nonstandard elements . . . .

By contraposition, we can also formulate the important principle

> if all elements of a set E are standard, then E is finite.

This principle can—and will—be made even more precise in (2.4.2) below. For reference, we give a number to our observation.

1.4.4

(*Basic principle.*) *In any infinite set E, there are nonstandard elements.*

1.5 MORE COMMENTS SUGGESTED BY THE EXAMPLES

1.5.1 Set-forming relations

The application (1.4.2) has revealed a finiteness character of the notion of standard element. Indeed, it is always possible to find a finite subset $F \subset E$ containing all standard elements. However, we should not conclude that 'the subset of standard elements of E is finite'. The property 'x is standard' is not set forming . . . . Indeed, it is time to recall that not all properties $P(x)$ are **set forming**. It is not enough to write $\{x : P(x)\}$ to prove the existence of a set E containing exactly the elements x for which $P(x)$ holds. For example, the property '$x = x$' is not set forming: there is no set containing all sets x. There is no universe. A famous paradox arises if we consider the property 'x is a catalogue which does not mention itself'. If we try to make a catalogue E of these catalogues x not mentioning themselves, we are faced with a dilemma: should we list E itself in the big catalogue? Both decisions lead to a contradiction.

Physicists experience this difficulty when they consider properties like $x \gg 1$: they would never consider the interval of all x's with this property. What would be its lower bound? The relation '$x \gg 1$' is not set forming and is very similar to the

'x is nonstandard' relation. For natural integers, 'n nonstandard' can very well be interpreted by '$n \gg 1$'.

Mathematicians will certainly experience some difficulty in working with these 'non set forming' relations! Indeed, this is one fundamental difference of attitude towards mathematics . . . and we shall have to come back to it with some insistence!

1.5.2 Axiom of specification

The ZF set theory avoids classical paradoxes by the following principle. To specify a set, it is necessary to start with a set. More explicitly, it postulates that if E is a given set and P is any property (applying to the elements of E), then there is a set E_P—a part of E—consisting precisely of the elements x of E for which $P(x)$ is true:

$$E_P = \{x \in E : P(x)\}.$$

In particular, the relation $P(x)$ given by '$x = x$' is perfectly legitimate and leads to $E_P = E$. In other words, to form a set—with ZF—, we have to start from a set. One gets nothing from nothing . . . .

1.5.3 A restriction of (IST)

From now on, we shall have to restrict the application of the specification axiom to classical properties (as defined in 1.2.2). If E is a set and P a classical property, we can still speak of the set $\{x \in E : P(x)\}$. Its existence is furnished axiomatically by (ZF) that is embedded in NSA. *But* if the property P contains explicitly or implicitly the 'standard' term of (IST), there is—in general—no subset of E containing exactly the elements x of E for which $P(x)$ holds. If it exists, it has to be proved by some method (e.g. by specifying it in another way). In other words, if a property P is nonclassical, before writing $\{x \in E : P(x)\}$, one has to prove that P *is* set forming. Typically, the relation 'x is standard' is in general not set forming. But of course, if $E = \varnothing$ is the empty set, the relation 'x is standard'—or any relation—*is* set forming and $\{x \in \varnothing : x \text{ is standard}\}$ is the empty set!

Similarly, if the set $E = \{a\}$ consists of precisely one standard element a, the property 'x is standard' is set forming in E and $\{x \in E : x \text{ is standard}\}$ is the set $E = \{a\}$ itself.

1.5.4 A warning

The restriction formulated above certainly deserves special attention. Thus we formulate it as explicitly as possible

ONLY CLASSICAL RELATIONS $P(x)$ ON A SET E SHOULD BE USED WITHOUT PRECAUTION TO FORM SUBSETS OF E.

We encounter here a first and main difference between (ZF) and (IST). Any subset which could previously be formed can still be constructed. But the new properties, containing the attribute 'standard' in any disguised form, should not be used to define subsets.

For example, the set $\mathbb{N}$ of natural numbers is infinite. Hence, it contains nonstandard elements. More precisely, we have seen that it contains nonstandard elements x with $x > n$ for all standard integers $n \in \mathbb{N}$. But we should not infer that there is a smallest nonstandard integer . . . . It is true that any nonempty subset of $\mathbb{N}$ has a smallest element (this is a classical statement, hence still valid in the new theory). But there is no subset of $\mathbb{N}$ consisting precisely of the nonstandard integers. In fact, we shall see that if n is a nonstandard integer, $n-1$ is still a nonstandard integer. The other axioms of (IST) will however imply that the integers 0, 1, 2 (and others) are standard. There is no smallest nonstandard integer (there is no smallest $n \gg 1$).

The impossibility of using nonclassical properties to specify subsets *is at the root of the coherence of* (IST). Violating it immediately leads to contradictions.

1.5.5 The induction principle

Since the **induction principle** is based on the well ordering of the set $\mathbb{N}$ of natural integers

EVERY NONEMPTY SUBSET OF $\mathbb{N}$ HAS A SMALLEST ELEMENT,

it remains valid . . . provided we can define the subset in question. The property that specifies it should be classical. The induction principle thus takes the form

IF P IS A CLASSICAL PROPERTY FOR WHICH $P(0)$ IS TRUE AND SUCH THAT

$$P(n) \Rightarrow P(n+1) \quad (\text{FOR ALL } n \in \mathbb{N}),$$

THEN $P(n)$ IS TRUE FOR ALL $n \in \mathbb{N}$.

A nonstandard induction principle can also be established for *all* properties P. It will be studied as Exercise (2.8.4).

1.5.6 Finiteness comment

If E is a set, we cannot say that E has only finitely many standard elements since there is—in general—no subset of E containing only these elements. But there is a finite subset $F \subset E$ containing all standard elements of E, and usually many others!

This finiteness property should not be considered as shocking! Indeed, we should interpret the predicate 'standard' by 'accessible', or 'specifically observable', . . . and due to our limitations (in space, time, . . .) accessible elements in any set *should* be contained in a finite subset. An illustration of this phenomenon can be given with the set $\mathbb{R}$ of real numbers. In a computer, the set of real numbers is finite. But the size of this set varies with the type of computer. It also increases with each technological progress and is thus a function of time! With NSA, the notion of standard real number is a theoretical notion, independent of the people who use it, invariable in time.

1.6 NATURAL INTEGERS

1.6.1 Illimited integers

Let us come back to the integers n which are bigger than all standard integers. These nonstandard integers are the 'infinitely large' integers considered (!?) by Leibniz and Euler. We simply call them *illimited*. Indeed, the classical sentence

all natural numbers are finite,

is still true. Recall that in (ZF), all mathematical entities are sets: integers are sets. We have already mentioned that the integers 0 can be taken to be the empty set $\varnothing$. Similarly, it is usually understood that

$$1=\{\varnothing\}, \qquad 2=\{\varnothing,\{\varnothing\}\}=\{0, 1\}, \ldots .$$

One can take inductively $n+1=n\cup\{n\}$. Since all integers are finite sets, illimited integers are—*a fortiori*—finite sets.

We are not adding new elements to the classical set $\mathbb{N}$ but we are supposed to be able to distinguish the new property 'standard or not' among them. In complete analogy, if a result holds for all integers n, nobody should start having doubts when the particular integer under consideration is prime . . . .

1.6.2 Intervals of integers

We shall often denote by

$$I_n=[0, n[=\{m\in\mathbb{N}:m<n\}=\{0, 1, 2, \ldots, n-1\}$$

the initial interval of $\mathbb{N}$ containing the first n integers. By definition, (ZF) identifies this set with the number n itself, but it may be useful to consider it also as subset $I_n\subset\mathbb{N}$.

Since $n>m$ in $\mathbb{N}$ implies $n+1>m$, $n+2>m$, . . . and more generally $n+k>m$ (for all $k\in\mathbb{N}$), we conclude that if n is an illimited integer, all integers $n+k$ ($k\in\mathbb{N}$) are—*a fortiori*—illimited integers. This shows that illimited integers exist in abundance in $\mathbb{N}$! But let us resist the temptation of collecting them. Let us distinguish without taking apart . . . .

Of course, complements can be taken in set theory, and it is quite legitimate to consider

$$J_n = \mathbb{N} - I_n = \{m \in \mathbb{N} : m \geq n\} = [n, \infty[\text{ (also denoted } [n, \rightarrow[).$$

Continuing in this vein, we can argue

for each n, the interval I_n is a finite subset of $\mathbb{N}$,
but $\mathbb{N}$ is an infinite set,
hence J_n is an infinite set *containing only illimited integers.*

But of course, J_n does not contain all illimited integers: $n-1$ is still illimited but does not belong to J_n.

Similarly, if n is any illimited integer, the finite initial interval I_n contains all standard integers (i.e. all limited integers). It does contain illimited integers as well: $n-1$ is still illimited and belongs to I_n. This illustrates the general principle (1.4.2).

1.7 POWER SETS

In our first deductions using the idealization axiom (I), we have used the **Power set** $\mathscr{P}(E)$ of a set E. One axiom of set theory asserts indeed that

for any set E, there is a set E' with the property

$$A \in E' \Leftrightarrow A \subset E.$$

This power set E' is usually denoted by $\mathscr{P}(E)$: it is well defined by E since its elements are completely unambiguously defined when E is given. Since 'finite' is a classical notion (i.e. a classical property), the specification axiom allows us to define the subset

$$\mathscr{P}_f(E) = \{A \in \mathscr{P}(E) : A \text{ is finite}\}$$

of $\mathscr{P}(E)$. In the deduction (1.4.2) of (I), the classical binary relation

$R(A, y) =$ 'A is a finite subset containing the element y' was used. If we work in a reference set E, we can consider that R is defined on the Cartesian product $\mathscr{P}_f(E) \times E$ so that idealization produces an $x = F \in \mathscr{P}_f(E)$, i.e. a finite part F of E, containing all standard elements of E.

1.8 EXTERNAL SETS

It is true that mathematicians are so used to forming subsets with all kind of properties, that they may feel insecure about admitting nonclassical relations, since these are usually not set forming. Physicists will certainly not experience the same feeling since they constantly use relations of this type, a typical example being the famous $x \gg 1$ relation (in the set $\mathbb{N}$ of natural numbers, or in the set $\mathbb{R}$ of real numbers).

One way of dealing with this problem is to introduce a special notion of class for these 'nonsets'. They are currently called **external sets** (being understood that an external set is *not a set*!). Special care has to be taken when dealing with these external sets. For example, a nonempty external subset of $\mathbb{N}$ does not necessarily have a smallest element.

This way of treating nonclassical relations may have serious advantages and is similar to the von Neumann–Bernays approach to set theory where types are introduced. For example, in this set theory, there is a 'class' containing all sets. But in this theory, a set is never an element of another set. There is a distinction of type between elements and sets, sets and classes, etc.

1.9 EXERCISES

1.9.1 Take an illimited integer $n \in \mathbb{N}$. Can you show that any injective map of

$$I_n = \{m \in \mathbb{N} : m < n\}$$

into itself is surjective (i.e. onto)?

1.9.2 Give one or two examples of nonclassical properties which are nevertheless set forming (in a certain set E).

1.9.3 In this exercise, one can admit that the integers 1 and 2 are standard (this will follow from the axioms of next chapter). Consider the following classical statement (fundamental theorem of arithmetic):

> any integer $0 \neq n \in \mathbb{N}$ can be uniquely decomposed in a finite product of primes, say
>
> $$n = p_1^{v_1} \dots p_k^{v_k} \quad \text{with} \quad p_1 < \dots < p_k \quad \text{and} \quad v_i \in \mathbb{N}.$$

Taken n illimited.

(a) Can we say that all p_i are illimited?
(b) Can we say that the number k of prime factors of n is illimited?
(c) Can we say that at least one prime factor p_i of n is illimited?

(Idealization)

Le charme est rompu: l'illusion cesse.
(*Petit Robert*: dictionnaire)

[*When the charm is broken, the illusion ceases.*]

CHAPTER 2

STANDARDIZATION AND TRANSFER

2.1 AXIOMATIC: END

2.1.1 Standardization (S) and transfer (T)

The last two axioms (or axiom schemes) of NSA have consequences which are best examined together. Thus we formulate them together. The first one is the *S*tandardization axiom (S).

> Let E be a *standard* set and P any property (classical or not). Then there is a (unique) standard subset $A = A_P \subset E$ having for standard elements precisely the standard elements $x \in E$ satisfying $P(x)$.

More formally, under the stated assumptions,

$$\exists^s A \subset E \text{ such that } \forall^s x\ [x \in A \Leftrightarrow x \in E \text{ and } P(x)].$$

The second is the *T*ransfer axiom (T).

> If all parameters $A, B, \ldots, L$ of a *classical formula* F have *standard values*, then $F(x, A, B, \ldots, L)$ holds for all x's as soon as it holds for all standard x's.

Again, under the stated conditions, this means that

$$\forall x\ F(x, A, B, \ldots, L) \Leftrightarrow \forall^s x\ F(x, A, B, \ldots, L).$$

To avoid theoretical questions linked to our use of dots in the expression $F(x, A, B, \ldots, L)$, we should explicitly limit the number of parameters in such a formula to a thousand (or a million!) to fix ideas. Certainly, we do not want to discuss this point here.

2.1.2 Terminological comment

We have used the terms *property*, *relation*, *formula* in the statement of the axioms (resp. in (S), (I), (T)). They are essentially synonymous terms. However, 'property' should suggest that the formula has *one* distinguished variable, 'relation' has at least *two* variables (playing a distinguished role) whereas 'formula' is a general term including up to a thousand (or a million) variables.

Nelson has introduced the abbreviation IST for the theory based on (ZFC) and the three axioms (I), (S), (T), also calling this theory *I*nternal *S*et *T*heory. This is the IST point of view of nonstandard analysis. In spite of our aversion for acronyms, we shall occasionally use them. . . .

2.1.3 Dual form of the transfer axiom

Let us explicitly formulate the equivalent dual form (T′) of transfer. It is obtained by replacing F by its negation F'. By (T) (for a classical F with standard values for all parameters), the following are equivalent

for all x, $F(x, A, B, \ldots L)$ holds,
for all standard x, $F(x, A, B, \ldots, L)$ holds.

We can reformulate them respectively by

there is no x such that $F(x, A, B, \ldots, L)$ is wrong,
there is no standard x such that $F(x, A, B, \ldots, L)$ is wrong.

But '$F(x, A, \ldots, L)$ is wrong' is equivalent to 'F' $(x, A, \ldots, L)$ holds' and we obtain the dual form of (T) as equivalence between

there is an x such that $F'(x, A, B, \ldots L)$, and
there is a standard x such that $F'(x, A, B, \ldots, L)$.

Still for classical formulas in which all parameters have standard values, we can rewrite the transfer axiom in the form (T′)

$$\exists x \quad \text{with } F'(x, A, B, \ldots, L) \Leftrightarrow \exists^s x \quad \text{with } F'(x, A, B, \ldots, L).$$

From this formulation of the transfer axiom it follows immediately that all concepts which are well defined (i.e. uniquely defined) within classical mathematics *are* standard. Indeed, if there is a unique x such that $F(x, A, B, \ldots, L)$ (F classical, $A, \ldots, L$ having standard values), then *this* x must be standard.

Many examples of this situation will be considered in 2.2.

2.2 CONSEQUENCES

2.2.1 The sets

$$0 = \varnothing,\ 1,\ 2,\ 12^3,\ e,\ \pi,\ \mathbb{N},\ \mathbb{R},\ \mathbb{C},\ \mathbb{F}_2 = \mathbb{Z}/2\mathbb{Z}$$

are uniquely defined by means of classical formulas. Consequently, they are standard objects. For example, the number π can be defined by the following formula: it is the smallest positive zero of the sine function (indeed certain books take this for definition of π!). In general, we shall not even bother to give an explicit form of the classical expression defining a classical object . . . being content with the knowledge that such an expression indeed exists!

In particular, transfer in its dual form (T′) implies that all numbers defined explicitly and uniquely by a classical formula are standard. Thus, 3, 4 and -1 are standard, . . . But there exist nonstandard integers by (1.4.1). These nonstandard integers have a certain *charm* that prevents us from really grasping them! Their existence is axiomatic.

At this point, the reader may feel frustrated (?), but let us emphasize that all mathematics are built on abstractions. Who has actually *seen* an infinite set? (Nevertheless, we are accustomed to speak about the set $\mathbb{N}$ of natural integers.) Who has really seen the number '1'? (We can only see *incarnations* of the abstract notion of an integer . . . .) It would be completely wrong to think that since nonstandard integers only exist axiomatically, we shall never be able to make precise statements about them. On the contrary, some nonstandard integers are even, some are odd, some admit the prime divisor 3, some are congruent to 5 mod 7, etc. and it is our purpose to develop some confidence in working with them. (The exercises of the first chapter should already have conveyed that impression!)

2.2.2

Here is another example of a classical formula (with parameters) to which transfer can be applied:

> $a \neq 0$ and b are elements of a field k.
> and $x \in k$ satisfies $ax - b = 0$.

The parameters of the preceding statement are a, b, k and should be standard, if transfer is to be applied (take for example $k = \mathbb{Q}$ or $\mathbb{R}$ or $\mathbb{C}$). Thus we get

> if the field k is standard, $x = b/a$ is standard as soon as its numerator b and its denominator a are both standard.

2.2.3

Other classical formulas with parameters abound. Take for example

$$x = E \cup F, \quad x = E \cap F, \qquad x = E \times F, \quad x = F^E.$$

If the parameters E and F are standard, they define a standard x. Thus,

$$E \text{ and } F \text{ standard} \Rightarrow E \cup F, \ldots, \mathscr{F}(E, F) = F^E \text{ standard.}$$

Similarly,

$\mathscr{P}(E)$ is standard if E is standard

$I_n = \{m \in \mathbb{N} : m < n\}$ is standard if n is standard (actually, this is a tautology!)

$[a, b] = \{x \in \mathbb{R}: a \leq x \leq b\}$ $(\subset \mathbb{R})$ is standard if a and b are standard.

2.2.4

In its direct form, transfer (T) can be applied to the classical formula $F(x, E_1, E_2)$ given by '$x \in E_1 \Rightarrow x \in E_2$' if the two sets E_i (acting as parameters) are standard. In this case, we thus discover that

$$\forall x\ (x \in E_1 \Rightarrow x \in E_2) \quad \text{(namely } E_1 \subset E_2)$$

as soon as

$$\forall^s x\ (x \in E_1 \Rightarrow x \in E_2).$$

In words: to check that two standard sets E_1, E_2 verify the inclusion relation $E_1 \subset E_2$, it is sufficient to check that the standard elements x of E_1 belong to E_2. Any inclusion relation between two standard sets can be proved at the level of their standard elements.

The preceding argument also proves that two standard sets are equal as soon as they have the same standard elements. Set theory is based on the **extensionality axiom** asserting that two sets are equal exactly when they have the same elements. We have obtained a **transferred extensionality principle** valid for standard sets.

2.2.5

The uniqueness of the standard part A of E whose existence is postulated by (S) follows in fact from the preceding extensionality principle. Since we are not striving for minimal axioms, we have already mentioned uniqueness (between brackets) in the axiom itself. If E is a standard set, we can thus adopt a functional notation for this well defined standard part A corresponding to the property P. Nelson suggests that we use

$$A = {}^S\{x \in E : P(x)\}.$$

It is necessary to insist on the fact that the property P used in this construction does not have to be classical. Even if P is not set forming, the standard subset ${}^S\{x \in E : P(x)\}$ can be formed, but only the standard elements in this set are recognized by the property P. In general, A will contain elements x for which $P(x)$ is wrong (these elements of A violating P are necessarily nonstandard) and some elements (also nonstandard) satisfying P may fail to be in A!

However, standardization (S) is a good substitute for specification and set formation. We say that the property P **defines the subset** A **implicitly**.

2.3 AN EXAMPLE

Let E be a standard set and A any subset of E. We can consider the relation $P(x)$ given by '$x \in A$' and construct the standard part ${}^S A = {}^S\{x \in E : x \in A\}$. This standard part ${}^S A$ contains all standard elements that belong to A with no other standard element. For example, if $\nu \in \mathbb{N}$ is a nonstandard integer and the part $A \subset E = \mathbb{N}$ is given by $A = [0, \nu]$, ${}^S A$ must contain all standard integers (since they are all in A, being smaller than ν). There is only one standard part of $\mathbb{N}$ with this property, namely $\mathbb{N}$ itself: as explained in 2.2.4, the transfer axiom implies that the two standard sets ${}^S A$ and $\mathbb{N}$—having precisely the same standard elements—coincide. This proves

$${}^S[0, \nu] = \mathbb{N} \quad \text{for every nonstandard integer } \nu.$$

In a similar vein, let us consider the property $P(x)$ given by 'x is standard' in the set $\mathbb{N}$ of natural numbers. The standard subset

$$A = {}^S\{n \in \mathbb{N} : n \text{ is standard}\}$$

of $\mathbb{N}$ is perfectly well defined by (S). By definition, its standard elements are precisely the standard natural integers. As before, we must have $A = {}^S\{n \in \mathbb{N} : n \text{ is standard}\} = \mathbb{N}$.

One shows similarly that

$${}^S\{n \in \mathbb{N} : n \text{ is nonstandard}\} = \varnothing$$

since these two standard sets have the same standard elements!

2.4 STANDARD FINITE SETS

2.4.1 Back to a classical definition

The term **finite** is classically defined in set theory. A set E is reputed to be finite when either of the following two equivalent conditions is satisfied

(i) all injective maps $E \to E$ are onto,
(ii) E is equipotent to an initial interval $I_n = [0, n[$ of $\mathbb{N}$.

To say that E is equipotent to I_n means that there is a one-to-one map between E and I_n. In this case, the integer n is perfectly well defined (classically) and is called **cardinal** of E. By transfer (in its dual form (T′)) this number $n = \mathrm{Card}(E)$ is standard if E is a standard set.

But the cardinal of a set can very well be a standard integer even if this set is nonstandard. For example, if the set E consists of only one nonstandard element x

$$E = \{x\} \quad \text{is nonstandard (cf. Ex 2.8.1)}$$

but

$$\mathrm{Card}\,(E) = 1 \quad \text{is standard.}$$

2.4.2 Theorem

Let E be any set. The following two conditions are equivalent

(i) all elements $x \in E$ are standard,
(ii) E is standard and finite.

2.4.3 Proof

Let us start by giving a few equivalent conditions

$$\begin{aligned} \exists x \in E,\ x \text{ nonstandard} &\Leftrightarrow \exists x \in E, \quad \forall^s y (x \neq y) \\ &\Leftrightarrow \forall^{sf} F, \quad \exists x \in E \quad (x \neq y \text{ for all } y \in F) \\ &\text{(I)} \\ &\Leftrightarrow \forall^{sf} F, \quad \exists x \in E \quad (x \notin F) \\ &\Leftrightarrow \forall^{sf} F, \quad E \text{ is not contained in } F. \end{aligned}$$

Consequently, by negation we have

$$\forall x \in E,\ x \text{ is standard} \Leftrightarrow \begin{cases} E \text{ is contained in a} \\ \text{standard and finite set } F \end{cases}$$

In the sequence of this proof, this equivalence will be refered to by the siglum (*).
Now, if E is standard and finite, (*)$\Leftarrow$ can be applied with $F = E$ and this proves (ii)$\Rightarrow$(i).
Conversely, if all elements of E are standard, using (*), we can take a standard and finite set F containing E. But

$$\begin{aligned} F \text{ standard} &\Rightarrow \mathscr{P}(F) \quad \text{standard} \quad \text{(using (T'))}, \\ F \text{ finite} &\Rightarrow \mathscr{P}(F) \quad \text{finite}. \end{aligned}$$

Since E is an *element* of the standard and finite set $\mathscr{P}(F)$, the implication (ii)$\Rightarrow$(i) (already proved) shows that E is *standard*. Moreover, $E \subset F$ finite shows that E is finite. This completes the proof of (i)$\Rightarrow$(ii). (Isn't it amazing to have to rely on the implication (ii) $\Rightarrow$(i) to prove (i)$\Rightarrow$(ii)!)

2.4.4 An important principle

Let E be a standard infinite set, A a subset containing all standard elements of E. Then A does contain some nonstandard elements.
Indeed, assume that E is a standard set and the subset A contains all standard elements of E. If A contains only standard elements, it is standard and finite by (2.4.2). In particular, $E - A$ is also standard and

$E - A = \varnothing$ (both standard sets contain no standard element!).

Thus $E = A$ is finite in this case, contrary to our assumption.

2.5 FUNCTIONS AND GRAPHS

2.5.1 A result on couples and projections

A couple (a, b) is standard precisely when its two components a and b are standard. Similarly, an n-tuple $(a_1, \ldots, a_n)$ is standard precisely when n and all its components a_i are standard.

A map $f: E \to F$ is standard precisely when the sets E, F and the graph $\mathrm{Gr}(f)$ are standard sets.

2.5.2 Justification

The characterization of standard couples follows from the fact that the first and second components of a couple are uniquely characterized classically (and in fact, axiomatically in many set theories). Transfer (T′) (as explained in 2.1.3) immediately gives

$$(a, b) \text{ standard} \Rightarrow a \text{ and } b \text{ standard.}$$

Since the couple (a, b) is also completely uniquely determined classically by its two components a and b, we also have

$$a \text{ and } b \text{ standard} \Rightarrow (a, b) \text{ standard.}$$

The other characterizations mentioned in 2.5.1 also follow from (T′). Observe that the number n of components of an n-tuple is also characterized uniquely classically. As a consequence, the n-tuple $(1, \ldots 1)$ can be standard only if its number of components n is a standard integer!

Although the preceding justification is sufficient to give a proof of the properties stated in the result, a curious reader might be interested to know how (ZF) proceeds precisely with maps, couples, etc. Let us recall that all objects of (ZF) are sets, and in particular, **couples** are sets. The definition (as suggested by Kuratowski) is the following

$$(a, b) = \{\{a\}, \{a, b\}\} = \{x, y\}.$$

The only relevant point of this definition is the fact that the couple completely determines its first component and its second component, and conversely. Triples are treated similarly, e.g. by defining $(a, b, c) = ((a, b), c)$. Induction produces n-tuples for all $n \in \mathbb{N}$. **Cartesian products** are thus well defined.

Similarly, in (ZF), a map $f: \mathrm{E} \to F$ is identified to the triple $(E, F, Gr(f))$. By definition, the **graph** of f is the subset

$$\mathrm{Gr}(f) = \{(x, y) \in E \times F : y = f(x)\} \subset E \times F.$$

For example, if E and F are standard sets, the set $E \times F$ is standard and the first projection $E \times F \to E$ is a standard map. In the same vein, if (E_i) is a standard family of sets (i.e. the map $i \mapsto E_i$ is standard) all projections $p_j: \Pi E_i \to E_j$ are standard.

2.5.3 Properties of standard maps

Let f be a standard map $E \to F$ and let x be a standard element of E. Then, the intersection of the two standard parts $\{x\} \times F$ and $\mathrm{Gr}(f)$ consists of only one point $(x, f(x))$ of $E \times F$. Thus this point is standard (use (2.4.2) since we only know—*a priori*—that the set $\{(x, f(x))\}$ is standard!). Consequently, $f(x)$ is standard: we have obtained

$$f \text{ and } x \text{ standard} \Rightarrow f(x) \text{ standard.}$$

Alternatively, one could also say that $f(x)$ is the unique element $y \in F$ satisfying $y = f(x)$. In this classical formula, the two parameters f and x are supposed to have standard values. Transfer furnishes y standard.

Another important consequence of transfer for standard maps is the following. Two standard maps $f, g: E \to F$ taking same values at all standard elements (of E) are equal.

Observe finally that a standard map can very well take nonstandard values (necessarily for nonstandard values of the variable!). For example, the identity map $f: \mathbb{N} \to \mathbb{N}$ is standard and we obviously have $f(n)$ standard exactly when n is standard. On the other hand, a standard map can also take standard values at nonstandard points (the constant map of $\mathbb{N}$ into itself, taking the value 0, has this property).

2.5.4 Implicit definition of a standard map

The subsets G in a Cartesian product $E \times F$ of two sets which are functional graphs are characterized by the property

$$\forall x \in E, \quad G \cap \{x\} \times F \text{ consists of precisely one element.}$$

When E, F and G are standard, this condition is satisfied as soon as it is satisfied for all standard x (Transfer). This simple fact allows one to define some standard maps in the following way. Suppose that a relation R (classical or not), given between two standard sets E and F, has the property

> for each standard $x \in E$, there is a unique $y \in F$
> such that $R(x, y)$ and this y is standard.

Then we can form the subset

$$G = {}^S\{(x, y) \in E \times F : R(x, y)\}.$$

It is standard by definition (axiom (S)) and is a functional graph since

$$\text{Card } [G \cap \{a\} \times F] = 1$$

for standard a first (hypothesis) and then also for all a by transfer. Don't forget to check that the parameters G and F of this formula have standard values (the number 1 can either be considered as standard value for an extra parameter, or as a constant of our theory. . .). In this case, we say that the relation R **defines implicitly the map** f. But one should be careful about the fact that—if R is not classical—$R(x, f(x))$ is not always true, and conversely, if $R(x, y)$ is true, we should not infer that $y = f(x)$. But it is true that if x is standard, $y = f(x)$ is characterized by the fact that $R(x, y)$ holds. For reference, we formulate as a principle these consequences of the axioms.

2.5.5 Principle of implicit definition of maps

Let E and F be two standard sets. Assume that a construction (classical or not) allows to define for each standard $x \in E$ a well defined standard element $y_x \in F$. Then, there is a unique standard map $f: E \to F$ such that $f(x) = y_x$ for all standard x.

In particular, if a construction (classical or not) gives, for each standard integer n, a standard element a_n (of some fixed standard set E), then there is a unique standard sequence $(a_n)_{n \in \mathbb{N}}$ taking the prescribed values for standard $n \in \mathbb{N}$.

2.6 RELATIVIZATION

A classical statement can always be **relativized** to standard sets as follows. Let us explain the procedure in a typical example without too much playing with quantifiers. Take a statement having the form

$$\text{for all } x \text{, there is a } y \text{ with } R(x, y)$$

where R is a classical relation. Let us introduce a property P by the abbreviation

$$P(x) = \text{'there is a } y \text{ with } R(x, y)\text{'}.$$

Since R is classical, P is also classical and transfer can be applied

$$P(x) \text{ holds for all } x \Leftrightarrow P(x) \text{ holds for all standard } x.$$

But now, if x is a fixed standard element, it is a standard value for the corresponding variable and we also have by transfer

$$\text{there is a } y \text{ with } R(x, y) \Leftrightarrow \begin{cases} \text{there is a standard } y \\ \text{such that } R(x, y) \end{cases}.$$

Altogether, we see that the properties

(i) for all x, there is a y with $R(x, y)$,
(ii) for all standard x, there is a standard y with $R(x, y)$,

are equivalent. The second statement is called a **relativized statement**. To relativize a classical statement, one has to replace all quantifiers $\forall$ appearing in it by their relativization $\forall^s$ and similarly all quantifiers $\exists$ by their relativization $\exists^s$.

We have said repeatedly that the NSA that we are considering knows of no enlargement. But at this point, it is possible to throw a bridge between the two points of view. One could also say that the classical statements have only been formulated (and proved) in their relativized version, considering that the purpose of transfer is to extend the classical results holding for standard elements only, to new, nonstandard elements. Eventually, we see that the two points of view present little psychological difference.

2.7 FINAL COMMENTS ON AXIOMATICS

2.7.1 Some difficult questions

It is true that if a set (or function, or any mathematical entity) is given, it can be either standard or not. The axioms of NSA should precisely enable us to discover which of the two

$$E \text{ is standard} \quad \text{or} \quad E \text{ is not standard}$$

is true. But it will not always be easy to discern which term of this alternative is the right one.

The situation should be compared to the classical question of discovering if a given set is finite or not. Here are two famous examples

- Let E be the set of prime numbers p such that the equation

  $$x^p + y^p = z^p \text{ has rational solutions with } xy \neq 0;$$

  is E finite or not?
- Let E be the set of complex numbers s with

  $$\zeta(s) = 0 \text{ and } 0 < \text{Re}\,(s) < 1,\ \text{Re}\,(s) \neq 1/2;$$

  is E finite or not?

Mathematicians conjecture that these two sets are in fact empty, but are not able to prove that they are finite! We hope to get an answer to these questions. . . .

2.7.2 Undecidable questions

Even if NSA should always allow one—in principle—to discover if a given set is standard or not, it is permitted to think that the question might be undecidable in some cases. More explicitly, a set E can be standard or nonstandard (excluded middle), but the theory may be too weak to prove either. We have experienced similar situations with the attribute 'finite' or 'empty'. For example is the set of cardinals $\mathfrak{a}$ with

$$\aleph_0 < \mathfrak{a} < \mathfrak{c}$$

finite (or even empty)? Here $\aleph_0 = \text{Card}(\mathbb{N})$ is the first infinite cardinal (the cardinal of countable sets) and $\mathfrak{c} = \text{Card}(\mathbb{R})$ is the continuum. P.-J. Cohen [C] has indeed proved that the answer to this question is independent from the axioms of set theory. In other words, either answer 'Yes' or 'No' can be added to classical set theory without introducing any contradiction.

However, transfer shows that all sets defined unambiguously classically are standard and this is a good start. . . .

2.7.3 A warning

Let us conclude this chapter with an explicit warning:

NEVER USE TRANSFER FOR AN EXPRESSION WHICH CONTAINS (EXPLICITLY OR IMPLICITLY) THE PREDICATE 'STANDARD' OR HAVING NONSTANDARD PARAMETER VALUES.

This warning, together with that given in (1.5.4) should allow beginners to discover the main reason for their first mistakes and thus resolve any contradiction that they will inevitably encounter (!?).

2.8 EXERCISES

2.8.1 Let E be a standard nonempty set. Show that E contains at least one standard element.

2.8.2 Let E be a standard infinite set. Show that there is an infinite subset C of E containing only nonstandard elements. Is there a standard subset C having the preceding properties?

2.8.3 Let E and F be two sets and assume that the Cartesian product $E \times F$ is standard. Can you prove that E and F are standard sets?

2.8.4 Prove the following **nonstandard induction principle**. Let P be any property (classical or not) with

- $P(0)$ is true,
- $\forall^s n \in \mathbb{N}, \quad P(n) \Rightarrow P(n+1)$.

Then $P(n)$ holds for all standard integers $n \in \mathbb{N}$.

2.8.5 Let n be an illimited natural integer ($n \in \mathbb{N}$ and n nonstandard) and $\varepsilon = 1/n \in \mathbb{Q}$. Compare the sets

$$A = \{X \in \mathbb{Q} : x < \varepsilon\},$$
$$B = {}^{S}\{x \in \mathbb{Q} : x < \varepsilon\},$$
$$C = \{x \in \mathbb{Q} : x \leqslant 0\}.$$

Which are standard?

2.8.6 Let R denote the relation on $\mathbb{N} \times \mathbb{N}$ defined on pairs of integers by

$$R(x, y) = \begin{cases} y = x & \text{if } x \text{ is standard} \\ y = x + 1 & \text{if } x \text{ is nonstandard.} \end{cases}$$

Show that this relation defines implicitly the identity function on $\mathbb{N}$.

2.8.7 Is there a function $f : \mathbb{N} \to \mathbb{N}$ with

$$f(n) = \begin{cases} n & \text{if } n \text{ is standard} \\ n + 1 & \text{if } n \text{ is nonstandard?} \end{cases}$$

Compare this situation with the preceding exercise.

2.8.8 (a) Let E be a set with Card(E) standard and finite. Show that the property $P(x) =$ 'x is standard' is set forming in E.

(b) Let E be a standard finite set. Show that all properties are set forming in E.

(c) Let E be a standard set containing a nonstandard element (hence E is infinite). Show that the set $E - \{x\}$ is nonstandard.

2.8.9 Let $x = a/b$ be a rational number expressed in reduced form i.e. $gcd(a, b) = 1$, $b > 0$ and $b = 1$ if $a = x = 0$. Prove that x is standard precisely when a and b are standard.

Ladies and gentlemen, all is perfect on the set...
... there are lots of things that aren't perfect in the set...
(Standardization)

Tout idéal, dès qu'il est formulé, prend un aspect désagréablement scolaire (V. Larbaud)

[*Any ideal, when clearly defined, becomes unattractive and bookish.*]

CHAPTER 3

REAL NUMBERS AND NUMERICAL FUNCTIONS

3.1 BASIC CONCEPTS

3.1.1 Notations

Starting from the set $\mathbb{N}$ of natural numbers, one constructs as usual the fundamental numerical sets

$\mathbb{Z}$: set of relative integers $\pm n$ ($n \in \mathbb{N}$),
$\mathbb{Q}$: field of rational numbers a/b ($a, b \in \mathbb{Z}$, $b \neq 0$),
$\mathbb{R}$: field of real numbers (completion of $\mathbb{Q}$),
$\mathbb{C}$: field of complex numbers (algebraic closure $\mathbb{R}(\sqrt{-1})$ of $\mathbb{R}$).

Everything proved classically for these numerical sets is still valid in our context. For example, the field $\mathbb{R}$ of real numbers is an **Archimedean field**:

> if a, b are two real numbers with $b > 0$, there always exists a positive integer $n \in \mathbb{N}$ such that $nb > a$.

3.1.2 Definitions

When x is a real (or complex) number, we shall say that x is **limited** if there exists a standard real number $y \in \mathbb{R}$ with $|x| \leq y$. We shall also say that x is **infinitesimal** (or **infinitely small**) if for every standard $y > 0$ we have $|x| < y$. When $x - y$ is infinitesimal, we say that x and y are **infinitely near** each other and we write $x \simeq y$. Due to the importance of these definitions, we display them in a more structured way

$$
\begin{array}{lll}
x \text{ limited} & \Leftrightarrow \exists^s y & \text{with } |x| \leq y \\
 & \Leftrightarrow \exists^s z & \text{with } |x| < z, \\
x \text{ infinitesimal} & \Leftrightarrow x \simeq 0 \Leftrightarrow \forall^s y > 0, \; |x| < y & \\
 & \Leftrightarrow \forall^s z > 0, \quad |x| \leq z, & \\
x \simeq y \; (x \text{ infinitely near } y) & \Leftrightarrow x - y \simeq 0 & \\
 & \Leftrightarrow \forall^s c > 0, \quad |x - y| < c. &
\end{array}
$$

An integer is limited exactly when it is standard. But if the integer n is nonstandard hence illimited, the rational $a = 1/n \in \mathbb{Q}$ is limited (e.g. $1/n \leq 1$) without being standard. Even more is true: the rational $1/n \simeq 0$ is infinitesimal when n is illimited.

To illustrate the use of this terminology let us make a few comments on the Archimedean property of real numbers. Take two strictly positive real numbers a and b. When b is limited and a illimited (i.e not limited, unlimited) in order to have $nb > a$, it is necessary to take n illimited. *A fortiori*, when b is infinitesimal $nb > a$ can only hold when n is illimited. But this necessary condition is by no means sufficient! For example, if n is illimited and $b = 1/n^2$, $nb = 1/n$ is still infinitesimal and is not bigger than $a = n$ or $a = 1$.

3.1.3 A few rules

Let us collect a few useful rules in dealing with the precedingly defined notions. First, if $a \neq 0$, it is obvious that

$$a \simeq 0 \text{ (i.e. } a \text{ infinitesimal)} \Leftrightarrow 1/a \text{ illimited.}$$

Second, if $a \simeq 0$ and b is limited, then $a \cdot b \simeq 0$. For example, if a is infinitesimal and n is a limited integer, we see that $n \cdot a$ is still infinitesimal.

Third, if n is a limited integer

$$a_i \simeq 0 (1 \leq i \leq n) \Rightarrow \sum_{1 \leq i \leq n} a_i \simeq 0.$$

More generally still if n is limited,

$$a_i \simeq b_i (1 \leq i \leq n) \Rightarrow \sum_{1 \leq i \leq n} a_i \simeq \sum_{1 \leq i \leq n} b_i.$$

Once correctly formulated, these assertions are immediately verified! Observe however that for n illimited, both implications would fail (take n illimited and $a_i = 1/n \simeq 0$ for all $1 \leq i \leq n$, so that $\Sigma a_i = 1$ is not $\simeq 0$). We shall need a generalization of these rules that we formulate separately.

3.1.4 Proposition (averages)

Let (a_k) and (b_k) $(1 \leq k \leq N)$ be two finite families of real (or complex) numbers with $a_k \simeq b_k$ for all integers k. Then

$$\sum_{1 \leq k \leq N} a_k/N \simeq \sum_{1 \leq k \leq N} b_k/N$$

(even if the number of terms N is illimited!).

3.1.5 Proof

If $\varepsilon > 0$ is standard, we certainly have $|a_k - b_k| < \varepsilon$ whence

$$\left| \sum_{1 \leq k \leq N} (a_k - b_k)/N \right| \leq \sum_{1 \leq k \leq N} |a_k - b_k|/N \leq N\varepsilon/N = \varepsilon.$$

This proves precisely that the difference of the two sums is infinitesimal. Observe that if N is limited, multiplication by N preserves the relation $\simeq$ and we find the preceding rules (3.1.3).

3.1.6 Integral parts of real numbers

The classical function $f(x) = [x] =$ **integral part** of x is well defined for $x \in \mathbb{R}$ and $x \geq 0$). It represents the biggest integer k with $k \leq x$. Since the field $\mathbb{R}$ is Archimedean, there indeed exists an integer k such that $k \cdot 1 = k > x$ and

$$[x] = \inf\{k \in \mathbb{N}: k > x\} - 1.$$

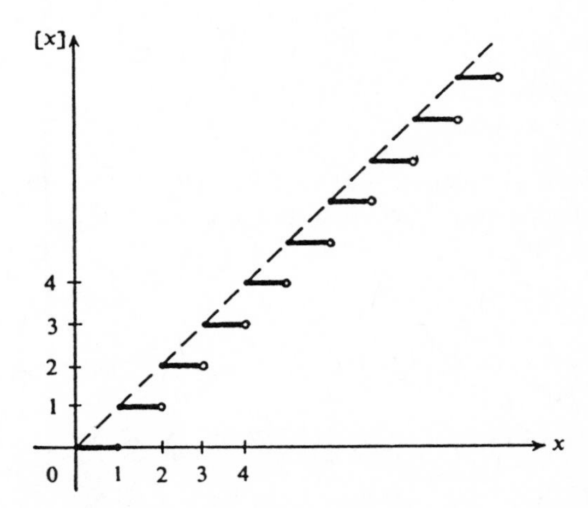

Fig. 3.1

By definition, we have the crucial inequalities

$$[x] \leq x < [x]+1, \quad 0 \leq x-[x] < 1, \quad |x-[x]| < 1.$$

We shall also consider functions of the type $f_\varepsilon(x)=[x/\varepsilon]\varepsilon$ where $\varepsilon>0$ is infinitesimal. These functions are not standard (cf. exercise 3.5.7). By definition

$$|f_\varepsilon(x)-x| < \varepsilon \quad \text{for all } x \geq 0$$

so that $f_\varepsilon(x) \simeq x$ for positive values of x. Figure 3.2 gives the graph of these functions (with a scale adapted to the situation!)

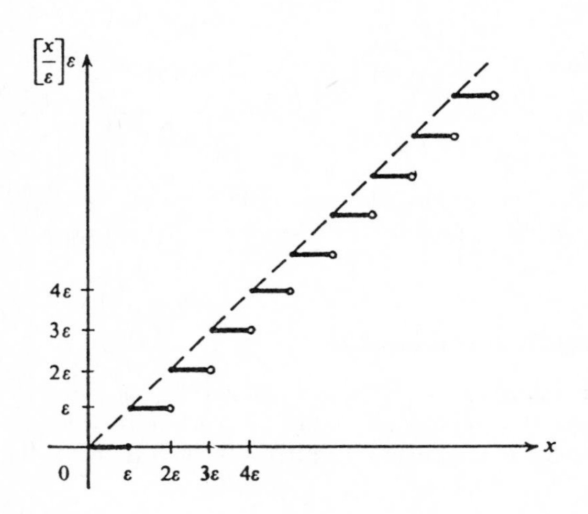

Fig. 3.2

3.2 STANDARD PART

3.2.1 Theorem–Definition

When $x \in \mathbb{R}$ is *limited*, there is a unique standard $x^* \in \mathbb{R}$ with $x \simeq x^*$. This real number x^* is called **standard part** of x and also denoted by $x^* = \mathrm{st}(x)$. It is only defined when x is limited.

3.2.2 Proof

Let us observe that the standard set ${}^S\{y \in \mathbb{R}\colon y \leq x\}$ is *bounded* so that we can define

$$x^* = \sup {}^S\{y \in \mathbb{R}\colon y \leq x\}.$$

(The reader will check that another equivalent possibility would be given by $x^* = \inf {}^S\{y \in \mathbb{R}\colon y \geq x\}$). In more detail, x being limited, there is a standard number

M with $|x| \leq M$. Let us compare the intervals

$$I = {}^S\{y \in \mathbb{R}: y \leq x\} \quad \text{and} \quad J = \{y \in \mathbb{R}: y \leq M\}.$$

If t is standard in I, we have

$$t \leq x \leq |x| \leq M$$

whence $t \in J$. Since both I an J are standard, this proves the inclusion $I \subset J$. In particular, I is *upper bounded* and we can let x^* be its upper bound $x^* = \sup I$. By transfer (T′), x^* is standard. To estimate $|x - x^*|$, let us choose any standard $a > 0$. We cannot have $x - x^* > a$ since this would imply

$$x^* < x^* + a < x$$

and x would not be a majorant of I ($x^* + a$ is standard and strictly smaller than x, hence in I). Similarly, we cannot have $x^* - x > a$ since this would imply

$$x < x^* - a < x^*$$

and $x^* - a$ would still majorize I (first $x^* - a > t$ for all standard $t \in I$ and transfer). But x^* was the smallest . . . . Thus, we have proved

$$|x^* - x| < a \quad \text{for all standard } a > 0$$

i.e. $x^* - x \simeq 0$ or $x^* \simeq x$. The *uniqueness* assertion results from the next lemma.

3.2.3 Lemma

There is a unique standard infinitesimal, namely 0.

3.2.4 Proof

If ε is a standard number, transfer is legitimate in the expression $\forall^s c > 0, |\varepsilon| < c$. Thus, if ε is standard and infinitesimal, we conclude that $|\varepsilon| < c$ for all $c > 0$. Obviously, this implies $\varepsilon = 0$ (if $\varepsilon \neq 0$, the choice $c = |\varepsilon| > 0$ would violate the required inequality).

3.2.5 Complex and vector case

Let $z = x + iy \in \mathbb{C}$ be a limited complex number ($i = \sqrt{-1}$). There exists a standard $M > 0$ with $|z| \leq M$ and thus

$$|x| \leq |z| \leq M, \qquad |y| \leq |z| \leq M.$$

This means that the real part $x = \mathrm{Re}(z)$ and the imaginary part $y = \mathrm{Im}(z)$ of z are limited. We can define the standard part of z by

$$z^* = x^* + iy^* (= \mathrm{st}(z)).$$

Obviously, similar definitions can be used in any vector space $\mathbb{R}^n$ or $\mathbb{C}^n$ . . . at least when the integer n is standard! A **limited vector** $x=(x_1, \ldots, x_n)$ is precisely an n-tuple with all components x_n limited (in $\mathbb{R}$ or $\mathbb{C}$). The **standard part** of a limited vector is the vector

$$\mathrm{st}(x)=x^*=(x_1^*, \ldots, x_n^*)$$

uniquely characterized by the properties

$$x^* \text{ is standard and } \|x-x^*\| \simeq 0.$$

Recall that the norm of a vector $x=(x_i)\in\mathbb{C}^n$ is defined by

$$\|x\|^2=\sum|x_i|^2$$

(it can also be applied to vectors with real components: $\mathbb{C}^n \supset \mathbb{R}^n$).

3.2.6 Some useful rules

The operation consisting in taking a standard part is the most important notion of NSA. It is useful to codify its use by a few simple rules. The following identities are immediately verified (when x and y are limited)

$$x\geq 0 \Rightarrow x^* \geq 0 \qquad (x\in\mathbb{R}),$$
$$(x+y)^*=x^*+y^*,$$
$$(x\cdot y)=x^*\cdot y^*.$$

In particular, for real numbers, we also have

$$x\geq y \Rightarrow x^* \geq y^*$$

(but observe that $x>y$ only implies $x^*\geq y^*$). We shall often use the last implication in the following form

x and y are standard, ε is infinitesimal

$$x\leq y+\varepsilon \Rightarrow x\leq y.$$

Since $(1+\varepsilon)^*=1$, strict inequalities are not preserved by the operation 'standard part'.

Finally, the reader can establish (cf. (3.1.3))

$$(x_1+\ \ldots\ +x_n)^*=x_1^*+\ \ldots\ +x_n^*$$

when all x_i's are limited *and* n is standard!

3.2.7 Theorem

Let $X\subset\mathbb{R}^n$(or $\mathbb{C}^n$) be a standard subset. Then X is compact if and only if

each $x\in X$ is limited and $x^*=\mathrm{st}(x)\in X$.

3.2.8 Proof

The classical characterization of compact subsets of $\mathbb{R}^n$ (or $\mathbb{C}^n$) is supposed to be known. It is:

X is closed and bounded.

Assume *first* that X (standard) is closed and bounded. Hence, there is a positive R with $\|x\| \leq R$ for all $x \in X$. Since X is a standard parameter, transfer (T′) is permitted and leads to

there is a standard positive R with $\|x\| \leq R$ for all $x \in X$.

In particular, all $x \in X$ are *limited.* Moreover, for each standard $\varepsilon > 0$, we have $\|x - x^*\| < \varepsilon$. Denote by $B_\varepsilon(x^*)$ the open ball of radius ε centred at the point x^*. Then

$$\forall^s \varepsilon > 0 \quad B_\varepsilon(x^*) \cap X \neq \varnothing$$

and by transfer (the parameters X, x^* are standard)

$$\forall \varepsilon > 0 \quad B_\varepsilon(x^*) \cap X \neq \varnothing.$$

Hence $x \in \bar{X} = X$ and we have proved the first part of the theorem.

Conversely, assume that all $x \in X$ are limited with $x^* \in X$. If we take any illimited $R > 0$, we shall certainly have $\|x\| < R$ for all $x \in X$. Hence X is contained in a ball of radius R, and is *bounded.* Moreover, if $y \in \bar{X}$, we can take $x \in X$ with $x \simeq y$ (take any point in the intersection of X and a ball centred at y and infinitesimal radius). By assumption $x^* \in X$ and thus $y \simeq x^*$ must be an equality if y is standard. This proves that all standard points y of the standard set $\bar{X}$ are in X: by transfer $\bar{X} \subset X$ and X is *closed.*

3.2.9 Standard part of a function

When $f\colon \mathbb{R} \to \mathbb{R}$ is a function such that $f(x)$ is limited for all standard x, it is possible to define a standard function f^* playing the role of **standard part of** f by

$$\text{Graph } (f^*) = {}^S\{(x, y) : y = f(x)^*\}.$$

Thus, the relation $y = f(x)^*$ implicitly defines f^*. In this relation, we could equally well replace the relation '$y = f(x)^*$'—only defined for the values of x for which $f(x)$ is limited—by the relation 'y is standard and $y \simeq f(x)$' well defined for all couples, possibly wrong for certain of them (e.g. wrong for all couples having a first component x fixed and such that $f(x)$ is not limited).

Observe that, in general, $f(x)$ and $f^*(x)$ are not infinitely close for all values of x. *A priori*, only standard values of x allow one to write $f^*(x) \simeq f(x)$ (when x is standard, we have precisely $f^*(x) = \text{st } f(x)$). For example, consider a fixed infinitesimal $\varepsilon \neq 0$ and define the function $f = f_\varepsilon$ by

$$f(x) = \begin{cases} 0 & \text{if } x \neq \varepsilon \\ 1 & \text{if } x = \varepsilon \end{cases}.$$

Then one checks that $f^* = 0$ so that $f^*(\varepsilon) - f(\varepsilon) = 1$ is not infinitesimal.

On the other hand, if f is equal to a limited constant c, then f^* is also constant and $f^* = c^*$.

3.2.10 Comments

The operation consisting in taking the standard part of real numbers is not a map. A first reason is that there is no set consisting of precisely the limited numbers. The domain of definition of a function which would consist in taking the standard part cannot even be formed. *Worse*: even if one restricts oneself to the interval $[-1, 1]$ containing only limited numbers, there is no map f—not even a nonstandard f—with $f(x) = x^*$ for all x in this interval (cf. exercise 3.5.6).

The classical situation 'x tends to y' (where x is variable and y fixed) is dissymmetric from the start. In NSA, it will often be replaced by a situation '$x \simeq y$ and y standard' having a first symmetric part. Equivalently, we can write $y = x^*$.

3.3 GENERALIZATION TO TOPOLOGICAL SPACES

3.3.1 Intrinsic definition of the standard part

As we defined it, the construction of a standard part depends on the possibility of using inequalities, e.g. to define what is meant by $x \simeq 0$ (for x real, complex or in $\mathbb{C}^n$). However, standard parts can be defined in a more general context.

As a motivation for the general definition, we observe that if x is a limited real number, its standard part $x^* \simeq x$ and x certainly belongs to all intervals $]x^* - \varepsilon, x^* + \varepsilon[$ where $\varepsilon > 0$ is a standard number. These intervals are standard neighbourhoods of x^* and we see that

any standard neighbourhood of x^* contains x.

In this form, the notion of standard part can be introduced in any topological space.

Let X be a topological space. Then, a point $x \in X$ is said to be **near standard** when there exists a standard $x^* \in X$ having the property

every standard neighbourhood of x^* (in X) contains x.

Thus, 'limited' is replaced by 'near standard' in an arbitrary topological space. To be still able to call x^* 'the' standard part of x when x is near standard, it is necessary to have uniqueness of this standard number. When the space X is

Hausdorff (distinct points have disjoint neighbourhoods, i.e. the space is (T_2)), it is obvious that this uniqueness condition is satisfied. In this case, for x near standard, we say that $x^* = \mathrm{st}(x)$ is the **standard part** of x.

3.3.2 Proposition

Let X be a standard Hausdorff space and $A \subset X$ a standard subset. If $x \in A$ is near standard (in X), then $x^* \in \bar{A}$.

3.3.3 Proof

If V is a standard neighbourhood of x^*, V contains $x \in A$ hence $V \cap A \neq \varnothing$. Transfer can now be used in the expression

for each standard neighbourhood V of x^*, $V \cap A \neq \varnothing$

and we find

for each neighbourhood V of x^*, $V \cap A \neq \varnothing$.

(The parameters of the expression in which we have used transfer were x^* and A, both standard.) This proves that x^* belongs to the closure of A as asserted.
Observe that transfer would have been illegal in the expression

for each standard neighbourhood V of x^*, $V \cap A \ni x$,

since this expression contains the nonstandard value x as parameter. Indeed, we see that—in general—x will not belong to all neighbourhoods of x^*.

3.3.4 Standard part of a set: shade

Let A be a subset of a (Hausdorff) topological space X. We can define a standard subset A^* of X by

$$\begin{aligned} A^* &= {}^S\{x \in X\text{: every standard neighbourhood of } x \text{ cuts } A\} \\ &= {}^S\{x \in X\text{: } \forall^s\ V \text{ neighbourhood of } x,\ V \cap A \neq \varnothing\}. \end{aligned}$$

For example, one checks that if $A = \{x\}$ consists of a single real point, $A^* = \varnothing$ if x is not limited whereas $A^* = \{x^*\}$ if x is limited.
In a metric space, the subset A^* can be implicitly characterized by the property 'there exists $y \in A$ with $d(x, y) \simeq 0$'.
This standard set A^* is also called **shade** of A and is sometimes defined in a slightly different (but equivalent) way (cf. Exercise 3.5.16 below). This standard set A^* is in general distinct from the standard set ${}^S A = {}^S\{x \in E\text{:} x \in A\}$ introduced in the preceding chapter (Section 2.3). For example, if $x \in \mathbb{R}$ is limited without being standard, ${}^S\{x\} = \varnothing$ but we have just seen that the shade of $\{x\}$ is $\{x^*\}$.

3.3.5 Proposition

The subset A^* of the standard metric space X is always closed.

If the subset A is standard, then $A^* = \bar{A}$ is the closure of A (and in particular $A \subset A^*$ in this case).

3.3.6 Proof

Since A^* is standard by definition, its closure is also standard and equality will follow from the fact that A^* and its closure have the same standard elements. Take x standard in the closure of A^*. There is a standard sequence (cf. Exercise 3.5.19) $(x_n) \subset A^*$ with $\text{dist}(x, x_n) < 1/2n$. But by definition, x_n is standard whenever n is, and for these, there is $y_n \in A$ with $x_n \simeq y_n$ so that *a fortiori* $\text{dist}(x, y_n) < 1/n$. We have thus shown

$$\forall^s n \quad \exists y_n \in A \quad \text{with } \text{dist}(x, y_n) < 1/n.$$

When the parameter A in this formula is standard (recall that we are assuming that x is standard!), transfer will give

$$\forall n \quad \exists y_n \in \mathrm{A} \quad \text{with } \text{dist}(x, y_n) < 1/n.$$

In this case, we see that $x \in \bar{A} = $ closure of A, and the second part of the proposition follows.

When A is arbitrary (nonstandard), transfer is no more applicable and it is necessary to proceed slightly differently. Look at the part

$$\{n \in \mathbb{N}\colon \exists y_n \in A \ \text{dist}(x, y_n) < 1/n\}.$$

As we have seen, this part contains all standard integers, hence contains some nonstandard integer ν (a set containing only standard elements is standard and finite). Thus

$$\exists y = y_\nu \in A \text{ with } \text{dist}(x, y) < 1/\nu \text{ infinitesimal,}$$

i.e. $x \simeq y \in A$ and $x \in A^*$ as was to be shown to prove the first part of the proposition.

3.4 SEQUENCES

3.4.1 Theorem

Let $(a_n)_{n \in \mathbb{N}}$ be a standard sequence of real numbers. Then, the following properties

(i) $a_n \to a$ for $n \to \infty$,
(ii) a is standard and $a_n \simeq a$ for all illimited $n \in \mathbb{N}$,
(iii) $\text{st}(a_n) = a$ for all illimited $n \in \mathbb{N}$.

3.4.2 Proof

A standard sequence is a standard map $A: n \mapsto a_n$. If it converges, its limit must be standard (by transfer (T')). By definition of convergence, for every standard $\varepsilon > 0$,

$$\exists N \text{ with } n \geq N \Rightarrow |a_n - a| < \varepsilon.$$

By transfer, we can even find a standard N with the same property. In this case, $n \geq N$ is automatically satisfied if n is illimited and we infer that

$$\forall^s \varepsilon > 0,\ |a_n - a| < \varepsilon \quad \text{for all illimited } n.$$

This means that $a_n \simeq a$ for all illimited n and proves (i)⇒(ii).

Observe that transfer was legitimate since the sequence was assumed to be standard. In detail, the classical statement

$$\exists N \text{ with } n \geq N \Rightarrow |a_n - a| < \varepsilon$$

can be rewritten

$$\text{the set of } n \text{ such that } |a_n - a| \geq \varepsilon \text{ is finite}$$

or even

$$A^{-1}\{x \in \mathbb{R}: |x - a| \geq \varepsilon\} \text{ is finite.}$$

In this form, all parameters are visible. They are A, a and ε (if we consider $\mathbb{R}$ as a constant, but we could equally well introduce an extra-parameter k taking any of the following standard values $\mathbb{R}$ or $\mathbb{C}$. . .).

(ii)⇔(iii) simply by definition of the standard part.

Finally, let us show that (ii)⇒(i). Thus we assume that for all illimited integer, $a_n \simeq a$. Let us take a standard $\varepsilon > 0$. Then

$$\exists N \text{ with } n > N \Rightarrow |a_n - a| < \varepsilon$$

is true (any choice of N illimited will do). This is a classical statement—having only standard parameters as before—true for all standard $\varepsilon > 0$. By transfer, it will remain true for all $\varepsilon > 0$ thereby proving convergence of the sequence to the limit a.

3.4.3 Corollary

Let (f_n) be a standard sequence of functions $I \to \mathbb{R}$ (where the interval I is standard), and f a standard function. Then, the following two properties are equivalent

(i) for each $x \in I$, $f_n(x) \to f(x)$ (namely, the sequence (f_n) converges simply, or pointwise, to f)
(ii) for each standard $x \in I$, $f_n(x) \simeq f(x)$ for all illimited integers n.

3.4.4 Proof

As in the proof of the theorem, let us stress that a standard sequence (f_n) is a standard map $F: n \mapsto f_n$ from $\mathbb{N}$ to some function space $\mathscr{F}$. Transfer shows that the two classical statements

for each $x \in I$, $f_n(x) \to f(x)$

and

for each standard $x \in I$, $f_n(x) \to f(x)$

are equivalent. But when x is standard, so are $a = f(x)$ and the sequence $n \mapsto a_n = f_n(x)$. The conclusion thus follows from the theorem.

3.4.5 Lemma

Let f: $I \to \mathbb{R}$ (or $\mathbb{C}$) be a function with $f(x) \simeq 0$ for all $x \in I$. Then f is bounded: more precisely $\sup |f(x)| \simeq 0$.

3.4.6 Proof

If $c > 0$ is standard, we have $|f(x)| < c$ for every x in I by assumption. Thus f is indeed bounded and $\sup |f(x)| \leq c$. Since c is arbitrary (standard and strictly positive) we get the conclusion.

3.4.7 Theorem

Let (f_n) be a standard sequence of functions $I \to \mathbb{R}$ (where the interval I is standard), and f a standard function. Then the following two properties are equivalent

(i) $f_n \to f$ uniformly on I,
(ii) for all $x \in I$, $f_n(x) \simeq f(x)$ for all illimited integers n.

3.4.8 Proof

The sequence $n \mapsto g_n = f_n - f$ is standard, and so is the sequence $n \mapsto a_n = \sup |g_n|$. Lemma (3.4.5) shows that Theorem (3.4.1) can be applied.

3.4.9 Robinson's lemma

Let $(a_n)_{n \in \mathbb{N}}$ be a sequence of real (or complex) numbers.

(a) If $a_n = 0$ for all standard n, there is an illimited $\nu \in \mathbb{N}$ with $a_n = 0$ for all $n \leq \nu$.
(b) If $a_n \simeq 0$ for all standard n, there is an illimited $\nu \in \mathbb{N}$ with $a_n \simeq 0$ for all $n \leq \nu$.

3.4.10 Proof

To prove (a), simply observe that the part

$$A=\{m\in\mathbb{N}\colon a_n=0 \text{ for all } n\leq m\}$$

contains all standard integers m, hence contains a nonstandard ν.

To prove (b), we cannot construct the same part A as above since the relation $a_n\simeq 0$ is not classical (illegal set formation) but Robinson's trick consists in forming the part

$$A=\{m\in\mathbb{N}\colon |a_n|<1/m \text{ for all } n\leq m\}.$$

Again, it is obvious that A contains all standard m since by assumption $a_n\simeq 0$ for all $n\leq m$. This set must contain an illimited integer ν for which $|a_n|\leq 1/\nu\simeq 0$.

3.5 EXERCISES

3.5.1 Show that there is an infinitesimal $\varepsilon>0$ and an integer k such that the finite set $\{0, \varepsilon, 2\varepsilon, 3\varepsilon, \ldots k\varepsilon\}$ contains all standard rationals in the interval $[0, 1]$.

3.5.2 Let $\varepsilon\neq 0$ be infinitesimal. Is there always an integer $n\neq 0$ with $n\varepsilon$ standard? Would your answer change if ε is assumed to be rational (use Exercise 2.8.9 for the characterization of standard rational numbers)?

3.5.3 (a) Construct an example of an unbounded function f such that $f(x)$ is infinitesimal for all limited $x\in\mathbb{R}$.

(b) Show that if $f(x)$ is infinitesimal for all limited x, then there is an illimited R with

$$f(x) \text{ infinitesimal for all } |x|=R.$$

(Continuous analogue of Robinson's lemma 3.4.9).

3.5.4 Let $f\colon \mathbb{R}\to\mathbb{R}$ be a map such that $f(x)$ is limited for all real x. Show that f is bounded and $\sup|f(x)|$ is limited.

3.5.5 Let $f\colon \mathbb{R}\to\mathbb{R}$ be a standard map. Show that the following properties are equivalent

(i) f is bounded,
(ii) there is a standard constant M with $|f(x)|\leq M$ $(\forall x\in\mathbb{R})$,
(iii) $f(x)$ is limited for all $x\in\mathbb{R}$.

3.5.6 Is there a (nonstandard) function $f\colon [0, 1]\to[0, 1]$ such that $f(x)=x^*$ for all $x\in[0, 1]$?

3.5.7 Show that the functions $f_\varepsilon(x)=[x/\varepsilon]\varepsilon$ are well defined for infinitesimal $\varepsilon>0$ (compare with the preceding exercise!). Are they standard?

3.5.8 What is the standard part of $\sqrt{(2+\varepsilon)}$ if ε is infinitesimal?

3.5.9 Give an example of a subset $A \subset \mathbb{R}$, $x \in A$, x limited but

$$x^* = \text{st}(x) \notin \bar{A}$$

(of course, 3.3.2 shows that one should take A nonstandard!).

3.5.10 Show that if n is an illimited (i.e. nonstandard) integer

$$e^x = \text{st}(1+x/n)^n \qquad \forall^s x \in \mathbb{R} \text{ (or } \mathbb{C}\text{)}.$$

3.5.11 Let $(a_n)_{n \in \mathbb{N}}$ be a sequence of real numbers and consider the following two properties

(i) $a_n \to a$ for $n \to \infty$,
(ii) a is standard and $a_n \simeq a$ for illimited $n \in \mathbb{N}$.

Find some examples proving that they are independent in general. [But recall that they are equivalent for standard sequences (a_n) by (3.4.1).]

3.5.12 Let (a_n) be a sequence which converges to a. Show that the sequence

$$s_n = (a_1 + \ldots + a_n)/n$$

also converges to a.

3.5.13 Let $x \in \mathbb{R}$ (or $\mathbb{C}$) be a number. Show that the following properties are equivalent

(i) x is limited and $x^* \neq 0$,
(ii) $x \neq 0$, x and $1/x$ are limited,
(iii) $x \neq 0$ and $\log|x|$ is limited.

A number satisfying them is called **appreciable**. Thus appreciable numbers are neither too small nor too large to be 'observed' (in fact, an actual measure of x would rather furnish the standard number x^*).

3.5.14 Let us define an equivalence relation $x \approx y$ (for couples with $xy \neq 0$) by requiring $x/y \simeq 1$. Check that this defines indeed a symmetric, reflexive and transitive relation on (nonzero) real (or complex) numbers. We read 'x is **asymptotic** to y' when $x \approx y$.

(a) Show that if x is appreciable (cf. preceding exercise), then

$$x \approx y \quad \text{if and only if } x \simeq y.$$

(b) Show that for two (finite or) infinite sequences (a_i) and (b_i)

$$a_i \approx b_i \quad \text{for all } i \geq 0 \Rightarrow \sum_i a_i \approx \sum_i b_i.$$

(Private communication from E. Nelson [N].)

3.5.15 Let E be a set. A **galaxy** in E is an external set $\mathscr{A}$ defined by a relation '$f(x)$ is limited' where $f: E \to \mathbb{R}$ is a function. Prove that two disjoint galaxies $\mathscr{A}$ and $\mathscr{B}$ are always contained in two disjoint subsets A and B of E. Similarly, a **halo** of E is an external set $\mathscr{C}$ defined by a relation '$f(x)$ is infinitesimal' where $f: E \to \mathbb{R}$ is a function. Prove that two disjoint halos $\mathscr{C}$ and $\mathscr{D}$ are always contained in two disjoint subsets C and D of E. (**Separation principle** of Diener and van den Berg.)

3.5.16 Let A be a subset of a metric space E with metric d. Recall that the distance of a point $x \in E$ to A is defined by

$$d(x, A) = \inf \{d(x, a): a \in A\}.$$

The **halo** $\mathscr{H}(A)$ of A is defined by the relation '$d(x, A) \simeq 0$'. Thus, by definition, A^* is the standard subset of E having the same standard elements as $\mathscr{H}(A)$. In other words, the relation '$\mathscr{H}(x)$ cuts A' implicitly defines A^* (cf. (2.2.5)). Check all these statements carefully.

3.5.17 Consider two strictly positive numbers a and b and let $E \in \mathbb{R}^2$ be the ellipse of equation

$$x^2/a^2 + y^2/b^2 = 1.$$

What is the standard part E^* of E when a is standard?

3.5.18 From $C_0 = [0, 1]$, let us withdraw the open middle third to keep $C_1 = [0, 1/3] \cup [2/3, 1]$. Define inductively the compact set $C_{n+1} \subset C_n$ by removing the middle thirds of the segments constituting C_n. The intersection C of all C_n is a nonempty compact set called Cantor set. Moreover, C is uncountable and negligible for Lebesgue measure on the real line. What is the standard part C_n^* of C_n when n is illimited?

3.5.19 Let X be a standard topological space and A a standard subset. If $x \in \bar{A}$ is in the closure of A, show that there is a standard sequence $(a_n) \subset A$ with $a_n \to x$.

3.5.20 Let X be a standard metric space and $A \subset X$ any subset. Show that the subset

$$U_A = {}^S\{x \in X: \text{all } y \simeq x \text{ are in } A\}$$

is **open** in X (U_A is implicitly characterized by halo$(x) \subset A$ and is therefore called 'interior halo' of A).

In my standard hat...
there is ...
a rabbit !
Hence this rabbit is standard !
(Transfer)

CHAPTER 4

CONTINUITY

4.1 S-CONTINUITY

4.1.1 Definition

If $f\colon \mathbb{R} \to \mathbb{R}$ is a function and $x \in \mathbb{R}$, we say that f is **S-continuous** at x when

$$y \in \mathbb{R} \text{ and } y \simeq x \Rightarrow f(y) \simeq f(x).$$

4.1.2 Theorem

Let $f\colon \mathbb{R} \to \mathbb{R}$ and $x \in \mathbb{R}$ both be standard. Then

$$f \text{ is continuous at } x \Leftrightarrow f \text{ is S-continuous at } x.$$

4.1.3 Comment

The notion of S-continuity is easy to manipulate from a logical point of view (there is no inversion of quantifiers). And at least for standard functions and points, continuity and S-continuity are equivalent. In a certain sense, the standard couples are the only ones which interest us explicitly. Even if we want to make universal statements on spaces of continuous functions and points, transfer will generally allow us to prove these by checking them on standard elements only. Thus we could ignore the notion of continuity and only define and use S-continuity. As we mentioned in the introduction, our purpose is primarily to explain NSA to people who are already accustomed to these classical notions. In particular, we shall study in detail the differences between continuity and S-continuity (for nonstandard functions and/or points). This attempt should not obscure the basic simplicity of the notion of S-continuity.

4.1.4 Proof of Theorem 4.1.2

All through the proof, we shall assume that the couple (f, x) is standard.

Suppose *first* that $y \simeq x \Rightarrow f(y) \simeq f(x)$. The following assertion is thus true for every standard $\varepsilon > 0$

$$\exists \delta > 0 \quad \text{with } \forall y\ (|y-x| < \delta \Rightarrow |f(y)-f(x)| < \varepsilon)$$

since it is sufficient to take any infinitesimal $\delta > 0$ to have

$$|y-x| < \delta \Rightarrow y \simeq x \Rightarrow f(y) \simeq f(x) \Rightarrow |f(y)-f(x)| < \varepsilon.$$

The preceding statement has the form

$$\forall^s \varepsilon > 0,\ P(\varepsilon)$$

with a classical property P in which the parameters f and x have standard values. Transfer is permitted and gives

$$\forall \varepsilon > 0,\ P(\varepsilon)$$

which is the classical definition of continuity of f at x.

Conversely, assume that f is continuous at x in the classical sense. Take $y \simeq x$ and try to evaluate the difference $f(y)-f(x)$. The classical definition of continuity gives in particular

$$\forall^s \varepsilon > 0 \quad \exists \delta > 0 \quad \forall y(|y-x| < \delta \Rightarrow |f(y)-f(x)| < \varepsilon).$$

Transfer (T′) can be applied to the inner part and allows to replace 'there exists a positive δ' by 'there exists a standard positive δ' (the parameters of the inner expression are f, x and ε all having standard values). Thus for standard $\varepsilon > 0$

$$\exists^s \delta > 0 \quad \forall y(|y-x| < \delta \Rightarrow |f(y)-f(x)| < \varepsilon).$$

Since we are assuming $y \simeq x$, the condition is automatically satisfied and we get

$$\forall^s \varepsilon > 0, \quad |f(y)-f(x)| < \varepsilon$$

which means $f(y) \simeq f(x)$. Thus the proof is completed.

4.2 EXAMPLES SHOWING THE DIFFERENCE BETWEEN CONTINUITY AND S-CONTINUITY

4.2.1

Let us consider the polynomial function $f(x) = x^2$. It is known that this function is continuous at *every* point (and this classical fact is still true!). Take now x illimited, so that $1/x$ is infinitesimal and $x + 1/x \simeq x$. We have

$$f(x+1/x) = (x+1/x)^2 = x^2 + 2 + 1/x^2 \simeq x^2 + 2$$

and for this point $x, f(x+1/x)$ is not infinitely close to $f(x)$:

$$\mathrm{st}(f(x+1/x)-f(x)) = 2 \ (\neq 0!).$$

4.2.2

Take now $a \neq 0$ and $f(x)=f_a(x)=a/(a^2+x^2)$. These functions are still continuous at all real points x. Let us show that if a is infinitesimal, they are not S-continuous at the origin $x=0$. Indeed, we have

$$f(0)=a/a^2=1/a \quad \text{illimited}$$

whereas

$$f(a)=1/(2a)$$

showing that

$$f(0)-f(a)=1/(2a) \quad \text{illimited although } a \simeq 0.$$

4.2.3

The classical function $\mathrm{sgn}(x)$ can be defined by

$$\mathrm{sgn}(x)=\begin{cases} 1 & \text{if } x>0 \\ 0 & \text{if } x=0 \\ -1 & \text{if } x<0 \end{cases}$$

Let us consider the functions $f_\varepsilon(x)=\varepsilon\cdot\mathrm{sgn}(x)$. They are discontinuous at the origin when $\varepsilon \neq 0$. But

$$|f_\varepsilon(y)-f_\varepsilon(x)| \leq 2\varepsilon \quad \text{for all couples } x \text{ and } y,$$

and thus, if ε is infinitesimal, f_ε is S-continuous at all points.

4.2.4

Let $\varepsilon>0$ be infinitesimal and define $f(x)=[x/\varepsilon]\varepsilon$ as in (3.1.6). By definition,

$$f(x)=k\varepsilon \quad \text{for } x\in[k\varepsilon, (k+1)\varepsilon[.$$

As we have

$$x-\varepsilon<f(x)\leq x \quad \text{for all } x\geq 0,$$

we infer

$$|f(y)-f(x)| \leq |y-x|+\varepsilon \quad \text{for all positive } x \text{ and } y.$$

This shows that f is S-continuous at all points. But f does 'jump' at all multiples of ε and is not continuous at these points.

This example is probably the best one to illustrate the difference in attitude with respect to (S-)continuity. Let us take a physical example. Continuous current in an electrical wire is classically represented by a continuous function. But more realistically, one should take into account the fact that current is produced by electrons. If measured very accurately at some precise point of the conductor, current would exhibit small jumps when individual electrons cross the measuring line. Thus, another possible modelization of continuous current would be given by a function of the type considered in this example. Since the elementary charge of the electron is *not* infinitesimal, it would again be an 'idealization' of reality. In a sense, one can even claim that S-continuity is better suited to our intuition than classical continuity.

4.3 RELATIONS BETWEEN CONTINUITY AND S-CONTINUITY

4.3.1 Standard continuous functions

By definition of S-continuity, we see that

$$\text{if } x \simeq y$$

then

$$f \text{ is S-continuous at } x \Leftrightarrow f \text{ is S-continuous at } y.$$

(Here, f can be standard or not.) But let us not try to gather in a set the points at which a given f is S-continuous (this property is not classical, hence not set forming in general). But if f is a standard function (defined in a standard interval $I \subset \mathbb{R}$), the two sets

$$\{x \in I: f \text{ is continuous at } x\},\ {}^S\{x \in I: f \text{ is S-continuous at } x\}$$

are both standard and have the same standard elements by Theorem (4.1.2). Thus, they *coincide*. In particular, for a standard function f (defined on some standard interval I), we have

$$f \text{ continuous on } I \Leftrightarrow \forall^s x \in I \quad f \text{ is S-continuous at } x.$$

4.3.2 A few rules

If f is S-continuous at a point x, the limited multiples of f will still be S-continuous at x

$$f \text{ S-continuous at } x \Rightarrow \forall^s a \in \mathbb{R},\ a \cdot f \text{ is S-continuous at } x.$$

Similarly, one observes that

$$f \text{ S-continuous at } x \text{ and } f(x) \text{ limited} \Rightarrow \forall^s n \in \mathbb{N},\ f^n \text{ is S-continuous at } x$$

(cf. Exercise 4.6.6).

4.3.3 Standard values of standard functions

Let f be a standard function which is S-continuous at a limited point x. Then

$$f(x^*)=f(x)^*.$$

Indeed, we can take the standard part of x since it is limited. The standard function f takes a standard value at the standard point x^*. Since f is S-continuous at x (or x^* by (4.3.1)) we have

$$f(x) \simeq f(x^*)$$

which gives the asserted equality by the uniqueness statement of (3.2.1). In particular, this shows that $f(x)$ is limited.

4.3.4 Implicit definition of continuity

We can still reformulate theorem (4.1.2) as follows. Let I be a standard interval and $\mathscr{F}$ the standard space of numerical functions on I. Consider the two sets

$$\{(f, x)\in\mathscr{F} \times I: f \text{ is continuous at } x\},$$

$$^S\{(f, x)\in\mathscr{F} \times I: f \text{ is S-continuous at } x\}.$$

The first is classically defined (with standard parameters I and $\mathscr{F}$). They are both standard and have the same standard elements, hence coincide. This is why we can say that S-continuity **implicitly defines** continuity.

It is not always easy to decide exactly which elements belong to a set of the form $^S\{\ \ldots\ \}$ since only the standard elements of such a set are—*a priori*— prescribed. But here (as in (4.3.1)), we have an exact classical description of these elements in special cases.

4.3.5 Theorem (continuous shade)

Let I be a standard interval and g an S-continuous numerical function on I. Assume that $g(x)$ is limited for all standard $x \in I$. Then the standard function f on I uniquely defined by $f(x)=g(x)^*$ for all standard $x \in I$ is continuous.

4.3.6 Proof

By transfer, it is enough to show that f is continuous at all standard points of I. Thus we shall consider and fix such a standard point $x \in I$.

By S-continuity of g, for each standard $\varepsilon > 0$, we have

$$|y-x| < \eta \Rightarrow |g(y)-g(x)| < \varepsilon$$

for all positive infinitesimal η. This proves that the *set* of all positive η for which

$$|g(y)-g(x)| < \varepsilon \quad \text{for all } y\in I \text{ with } |y-x| < \eta$$

contains all positive infinitesimal η. Thus this set must contain a standard $\eta > 0$ (there is no set consisting exactly of the infinitesimals . . .). Fix such a standard $\eta > 0$. Then it is obvious that for standard y

$$|y-x| < \eta \Rightarrow |f(x)-f(y)| \simeq |g(x)-g(y)| < \varepsilon$$

hence also $|f(x)-f(y)| < \varepsilon$ since both extreme terms are standard. To summarize, we have proved

$$\forall^s \varepsilon > 0, \quad \exists^s \eta > 0 \text{ with } \forall^s y, \quad |y-x| < \eta \Rightarrow |f(y)-f(x)| < \varepsilon.$$

A first transfer gives

$$\forall y \, |y-x| < \eta \Rightarrow |f(y)-f(x)| < \varepsilon$$

since the parameters in this classical expression are f, x, η and ε (all assumed to have standard values). In particular

$$\exists \eta > 0 \quad \text{with } \forall y \, |y-x| < \eta \Rightarrow |f(y)-f(x)| < \varepsilon,$$

again if the parameters f, x and ε are standard. A second transfer on ε will give

$$\forall \varepsilon > 0 \quad \exists \eta > 0 \quad \text{with } |y-x| < \eta \Rightarrow |f(y)-f(x)| < \varepsilon.$$

This proves continuity of f at x (both standard).

4.3.7 Examples

It is easy to illustrate the preceding theorem. Since the function $f: I \to \mathbb{R}$ (or $\mathbb{C}$) is assumed to be standard with

$$f(x) = g(x)^* \quad \text{for all standard } x \in I,$$

it means that $f = g^*$ (notation of (3.2.9)).

Take any standard continuous numerical function f on the standard interval $x \geq 0$ and choose an infinitesimal $\varepsilon > 0$. Then the function

$$g = f_\varepsilon \text{ defined by } g(x) = f([x/\varepsilon]\varepsilon) \quad \text{for } x \geq 0$$

is S-continuous with $g^* = f$ continuous.

4.4 UNIFORM CONTINUITY

4.4.1 Review of a classical definition

Uniform continuity of a function on a subset $J \subset I$ can be treated in a completely similar way as continuity at a point. As in 4.3, we still fix a standard interval I and denote by $\mathscr{F}$ the standard set of numerical functions on I. We define classically the standard set

$$\{(f, J) \in \mathcal{F} \times \mathcal{P}(I) \colon f|_J \quad \text{is uniformly continuous}\}.$$

Here, $f|_J$ denotes the restriction of the function f to J.

4.4.2 Theorem

When f and J are standard, $f|_J$ is uniformly continuous precisely, when

$$\forall (x, y) \in J \times J, \quad x \simeq y \Rightarrow f(x) \simeq f(y).$$

4.4.3 Proof

The given condition could be called S-uniform continuity, in complete analogy with (4.1.1) and this theorem is the analogue of (4.1.2).

Assume *first* that the condition is realized. Then, for each infinitesimal $\delta > 0$

$$|x - y| < \delta \Rightarrow x \simeq y \Rightarrow f(x) \simeq f(y) \Rightarrow |f(x) - f(y)| < \varepsilon$$

for all standard $\varepsilon > 0$. This proves

$$\forall^s \varepsilon > 0 \, \exists \delta > 0, \qquad (x, y) \in J \times J, \quad |x - y| < \delta \Rightarrow |f(x) - f(y)| < \varepsilon.$$

But the inner parameters of this classical formula are f and J, both standard, so that transfer can be applied. The conclusion is precisely the classical definition of uniform continuity.

Conversely, assume that $f|_J$ is uniformly continuous in the classical sense and take x, $y \in J$ with $x \simeq y$. we have to estimate the difference $f(x) - f(y)$ and show that it is infinitesimal. For each standard $\varepsilon > 0$, there is a standard $\delta > 0$ (by transfer) with

$$|x - y| < \delta \Rightarrow |f(x) - f(y)| < \varepsilon.$$

In particular, $x \simeq y$ will imply $|f(x) - f(y)| < \varepsilon$. As in (4.1.4), we conclude that when $x \simeq y$,

$$\forall^s \varepsilon > 0, \qquad |f(x) - f(y)| < \varepsilon,$$

and this proves $f(x) \simeq f(y)$.

4.4.4 Implicit definition of uniform continuity

Fix the standard interval (or subset) $I \subset \mathbb{R}$ and collect all uniformly continuous functions f on I

$$\{f \in \mathcal{F} : f \text{ is uniformly continuous}\}.$$

This set is classically defined (with a standard parameter I) hence is standard. Its standard elements are the same as the standard elements of

$$^S\{f \in \mathcal{F} : f \text{ is S-continuous at all } x \in I\}.$$

Hence these two sets coincide and uniform **continuity** is **implicitly defined** by

$$\forall x, y \in I, \qquad x \simeq y \Rightarrow f(x) \simeq f(y).$$

For a standard function f (defined on some standard I) we even have

$$f \text{ uniformly continuous on } I \Leftrightarrow f \text{ S-continuous at } \textit{all } x \in I.$$

It is good to realize that continuity of a function f, e.g. on the interval $]0, \infty[$ does not allow us to infer that

$$f(h) \simeq f(h') \quad \text{when } h \simeq h' \text{ are positive infinitesimals.}$$

Indeed, $h^* = h'^* = 0$ is not in the domain of definition of f. If $x \simeq y$ are illimited, we would not be able to say that $f(x) \simeq f(y)$ either. Both conclusions would however result from the uniform continuity of f.

4.5 THEOREMS ON CONTINUOUS FUNCTIONS

4.5.1 Theorem

Let $f: I \to \mathbb{R}$ (or $\mathbb{C}$) be a continuous function. Then, the restriction of f to any compact subset $J \subset I$ is uniformly continuous.

4.5.2 Proof

Assume first that f (hence I) and J are standard. We have to prove that

$$\forall x, y \in J, x \simeq y \Rightarrow f(x) \simeq f(y).$$

Since J is compact and standard, every point x of J is near standard, i.e. limited (explicitly, J is bounded, say $|x| \leq M$ for all $x \in J$, where we can assume M standard by transfer; then $x^* \in \bar{J} = J$ by (3.3.2)). We have $y \simeq x \simeq x^*$ and hence by S-continuity of f at the standard point x^*

$$\left.\begin{aligned} y \simeq x^* \Rightarrow f(y) \simeq f(x^*) \\ x \simeq x^* \Rightarrow f(x) \simeq f(x^*) \end{aligned}\right\} \Rightarrow f(y) \simeq f(x)$$

as expected.

Let us now turn to the general case and show how transfer is applicable. Denote by

$$C = {}^S\{(f, x) : f \text{ is S-continuous at } x\}$$

(observe that C is exactly the classical set consisting of couples (f, x) with f continuous at x) and

$$U = {}^S\{(f, J) : f|_J \quad \text{is S-continuous at all points } x \in J\}$$

(observe that U is exactly the classical set consisting of couples (f, J) with $f|_J$ uniformly continuous). These two sets C and U are standard and we have just proved that

$$\forall^s f, \forall^s \text{ compact } J, \qquad (f, x) \in C (\text{all } x \in I) \Rightarrow (f, J) \in U.$$

Since the two parameters C and U of this classical formula have standard values, transfer is legitimate and we infer

$$\forall f, \forall \text{ compact } J, \quad (f, x) \in C \text{ (all } x \in I) \Rightarrow (f, J) \in U.$$

4.5.3 Comments

The kind of transfer used in the preceding proof is so often used that it will usually be omitted. If necessary, the reader will be able to fill in the details in such situations. In particular, one should observe that transfer is legitimate without having to resort to the classical definitions of continuity and uniform continuity. The new (non classical definitions) enable us to form the standard sets C and U, permitting the use of transfer.

Let us also observe that the statement of (4.5.1) is classical. But the proof (4.5.2) is not classical. This is one typical aspect of nonstandard analysis: some classical statements are proved more economically with NSA.

The following theorem is also classical. Its proof emphasizes a typical relation between 'finite' and 'continuous' that is offered by NSA.

4.5.4 Theorem

Let $f: I \to \mathbb{R}$ be a continuous function. Then on each compact subset $K \subset I, f$ has a maximum.

4.5.5 Proof

Let $F \subset K$ be a finite part containing all standard points of this compact set. Let $a \in F$ be a point where the restriction $f|_F$ has its maximum (we are using the result of the theorem for the finite subset F: this is an easy classical result proved, e.g. by induction). Put $M = f(a)$ so that

$$f(x) \leq M \quad \text{for all } x \in F$$

and in particular, by choice of F

$$f(x) \leq M \quad \text{for all standard } x \in K.$$

To be able to apply transfer, we need to have standard parameters. Thus we assume f and K standard. Then, $f(x)$ is standard if x is, and we also have

$$f(x) \leq M^* = f(a)^* = f(a^*) \quad \text{for all standard } x \in K$$

(observe that a is limited since K is standard and bounded, and $a^* \in \bar{K} = K$). Transfer then gives

$$f(x) \leq M^* \quad \text{for all } x \in K.$$

Since $M^* = f(a^*)$, it is a maximum of f on K. We have thus proved that when f and K are standard, f has a maximum at a standard point. Transfer takes care of the general case as in (4.5.2).

4.5.6 Theorem

Let f: $I \to \mathbb{R}$ be a continuous function defined on some compact interval $I = [a, b]$ with $f(a) < 0$ and $f(b) > 0$. Then, there exist a point c, $a < c < b$ with $f(c) = 0$.

4.5.7 Proof

By transfer, it is still enough to prove this classical statement for standard data I and f. Let then F be a finite subset of I containing all standard points of this interval. Take for x the smallest element in the finite set F with $f(x) \geq 0$. Since a is standard, a is in F and by assumption $x > a$. The point x is limited and

$$f(x) \geq 0 \Rightarrow f(x^*) = f(x)^* \geq 0$$

(we have used the fact that f is standard through the equality $f(x^*) = f(x)^*$, cf. (4.3.3)). Let us still call y the largest element of the finite set F for which $y < x$:

$$y < x \quad \text{and} \quad f(y) < 0.$$

On the other hand, $y \simeq x$ since otherwise, $x - y$ would have a standard part $d > 0$, so that

$$y < y^* + d/2 < x \quad (\text{with } y^* + d/2 \text{ standard hence in } F)$$

would contradict the definition of x. By S-continuity of the standard function f at the standard point x^* we infer

$$0 > f(y) \simeq f(x^*) \simeq f(x) \geq 0.$$

Thus we can simply choose $c = x^*$.

In the case f standard, we have thus proved that there is a standard zero c in the (standard) interval I. But of course, f may also have nonstandard zeros (this can only happen when f has infinitely many zeros . . . since the set of zeroes of a standard f is automatically a standard set!).

4.6 EXERCISES

4.6.1 Discuss continuity and S-continuity for the following functions

(a) $f(x) = x^n$ where n is an illimited integer,
(b) $f(x) = \sin nx$ where n is an illimited integer

Show that there are some points at which the exponential e^x is not S-continuous.

4.6.2 Let f: $\mathbb{R} \to \mathbb{R}$ be a function satisfying

$$x \simeq y \quad \text{and} \quad x \leq y \Rightarrow f(x) \leq f(y)..$$

Show that f is monotonously increasing.

4.6.3 Give an example of a continuous function f: $\mathbb{R} \to \mathbb{R}$ such that

$$\begin{aligned} f(x) &= 1 \quad \text{for all standard rational numbers,} \\ f(x) &= 0 \quad \text{for all standard irrational numbers.} \end{aligned}$$

Show that this function is not S-continuous at limited points.

4.6.4 Let f: $I \to \mathbb{R}$ be a continuous function. Take an illimited integer n and put $\varepsilon = 1/n$. The function g defined on I by $g(x) = f([x/\varepsilon]\varepsilon)$ is a ruled function and is infinitely close to f at all points of I.

Is the preceding statement correct? (If not, correct it!)

4.6.5 Show that there exists an increasing function f: $\mathbb{R} \to \mathbb{R}$ which is discontinuous at all standard points. Construct such a function with $f(\mathbb{R}) \subset [0, 1]$.

4.6.6 Prove that the sum of two functions which are S-continuous at a point a in also S-continuous at that point. Can we say that the sum of a finite family of functions which are S-continuous at a point is S-continuous at that point? Is the product of two S-continuous functions at a point a also necessarily S-continuous at a?

4.6.7 Let us say that a function f: $\mathbb{R} \to \mathbb{R}$ is *suounitnoc* at a point $x \in \mathbb{R}$ when

$$\forall \varepsilon > 0 \qquad \exists \delta > 0 \qquad (|y - x| < \varepsilon \Rightarrow |f(y) - f(x)| < \delta).$$

This is a classical definition (in the sense of (1.2.2), i.e. it does not refer to NSA ... but I admit that it is not familiar!).

(a) Show that

$$f \text{ suounitnoc at a point} \Rightarrow f \text{ suounitnoc at all points.}$$

(b) Formulate the condition 'suounitnoc' in a more intuitive way (so that it would be easier to communicate the notion to a student ...).

(c) Observe that if f is standard

$$f \text{ suounitnoc at a point} \Rightarrow f(x) \text{ limited for all limited } x.$$

(This amusing exercise is taken from Nelson [9]. The definition reverses the roles of ε and δ in the continuity and in the spelling of this word . . . .)

$d - \epsilon \simeq d$

CHAPTER 5

DIFFERENTIABILITY

5.1 DIFFERENTIABLE FUNCTIONS

5.1.1. Difference quotients

Consider a function $f: I \to \mathbb{R}$ (or $\mathbb{C}$) defined on an interval of positive length and select an interior point $a \in I$. The function

$$g(x) = \frac{f(x) - f(a)}{x - a}$$

is well defined on $I - \{a\}$ and is continuous if f is continuous. Classically, we say that f is **differentiable** at the point a when g has a limit for $x \to a$. We keep this definition, but we are looking for another implicit definition, using NSA.

5.1.2 S-differentiability

Let f and a be as before. We shall say that f is **S-differentiable** if there exists a standard m with

$$\frac{f(x) - f(a)}{x - a} \simeq m \quad \text{for all } x \simeq a.$$

This notion is closely connected to usual differentiability as we shall presently see.

5.1.3 Theorem

Let $f: I \to \mathbb{R}$ (or $\mathbb{C}$) be a standard function and a be a standard interior point of I. Then

$$f \text{ is differentiable at } a \Leftrightarrow f \text{ is S-differentiable at } a.$$

5.1.4 Proof

The reader will have recognized the similarity with (4.1.2) identifying continuity and S-continuity for standard couples (f, a).

First assume that f is differentiable at a. Then

$$g(x)=\frac{f(x)-f(a)}{x-a}$$

has a continuous extension—still denoted g—with $m=g(a)$ $(=f'(a))$ standard since g is standard. Applying (4.1.2) to the standard couple (g, a) we get

$$g(x)\simeq g(\mathrm{a}) \quad \text{i.e.} \frac{f(x)-f(a)}{x-a}\simeq m \quad \text{whenever } x\simeq a.$$

This is the first implication given in the theorem.

Conversely, assume that f is S-differentiable at a, namely: there exists a standard m with $g(x)\simeq m$ for $x\simeq a$. Define $g(a)=m$ so that the extension—still denoted g—is standard and g is S-continuous at a. Since the couple (g, a) is standard, (4.1.2) can be applied and we infer that this extension of g is continuous at a. In other words, m is a limit for the difference quotients and f is differentiable at a with $f'(a)=m$.

5.1.5 Comments

We do not want to emphasize the distinction between differentiability and S-differentiability (in other words, we do not want to write a section parallel to (4.2)!). On the contrary, we prefer to consider that **S-differentiability implicitly characterizes differentiability**. We have indeed an identity of sets

$$\{(f, a): f \text{ differentiable at } a\}={}^{\mathrm{s}}\{(f, a): f \text{ S-differentiable at } a\}$$

(once more because both are standard and have the same standard elements. . . .). Ignoring the classical definition of differentiability, we see that we could *define*

$$f \text{ differentiable at } a$$

by

$$(f, a) \text{ belongs to } {}^{S}\{(f, a): f \text{ S-differentiable at } a\}.$$

In any case, we see that if f is standard, the set of points a at which f is differentiable is standard too, and the reader will check that if the interval I is standard, the set of all functions f which are differentiable at all points of I is also a standard set. . . .

5.1.6 An example

Consider the standard function $f(x) = \log x$ defined on the real positive axis $x > 0$. For $a > 0$ and $x = a + h$ with h infinitesimal, we have

$$f(a+h) - f(a) = \log(a+h) - \log a = \log\frac{a+h}{a}$$
$$= \log(1 + h/a) = h/a + h^2(\ldots).$$

From this expression, we see that for a standard first, f is (S-) differentiable at a with

$$f'(a) = 1/a.$$

Since the set of points where this standard function is differentiable is standard and contains all standard points $a > 0$, we infer that $f = \log$ is differentiable at all points $a > 0$. Moreover, since f must be standard whenever f is standard, we must have $f'(a) = 1/a$ at *all* points $a > 0$ for the logarithmic function.

5.1.7 Derivation rules

Let f and g be S-differentiable functions at a point a. Then,—for f, g and a standard first—it is obvious that the following relations hold

$$(f+g)'(a) = f'(a) + g'(a),$$
$$(fg)'(a) = f'(a)g(a) + f(a)g'(a),$$
$$(f/g)'(a) = [f'(a)g(a) - f(a)g'(a)]/g(a)^2.$$

Here, we have denoted by $f'(a)$ the standard number m whose existence is postulated in the definition of S-differentiability (and similarly for $g'(a)$).

When f and g can be composed, with g S-differentiable at a and f S-differentiable at $g(a)$, we also have

$$(f \circ g)'(a) = f'(g(a)) \cdot g'(a).$$

All these are easily established with differential quotients having infinitesimal numerators and denominators, giving 'infinitely good approximations' for derivatives as required in the definition of S-differentiability.

Since usual differentiability is implicitly defined by S-differentiability, the same relations hold for differentiable functions.

5.1.8 More comments

In principle, the notion of S-differentiability is completely sufficient for applications. But, for a classical mind, a perfect identification of the set (as given in (5.1.5))

$${}^S\{f, a) : f \text{ S-differentiable at } a\}$$

may be interesting. Technically, we shall have to use the following result, relating derivative (at a nonstandard point) and infinitesimal difference quotients (at the same point).

5.1.9 Proposition

Let $f: I + \mathbb{R}$ (or $\mathbb{C}$) be a standard function which is differentiable at the interior point x (standard or not) of I. Then, there exists a $\delta > 0$ with

$$\frac{f(y)-f(x)}{y-x} \simeq f'(x) \quad \text{for } 0 < |y-x| < \delta.$$

(Observe that $y \simeq x$ would usually not be sufficient to have 'proximity' of the differential quotient and the derivative!)

5.1.10 Proof

By assumption, the (standard) set of points where f is differentiable is not empty (it contains x). Hence this set J contains some standard points. Take any standard $a \in J$. Then, choosing an infinitesimal $\delta > 0$, we see that for a fixed standard $\varepsilon > 0$

$$0 < |y-x| < \delta \Rightarrow x \neq y \quad \text{and} \quad x \simeq y$$

$$\Rightarrow \frac{f(y)-f(x)}{y-x} \simeq f'(x)$$

$$\Rightarrow \left| \frac{f(y)-f(x)}{y-x} - f'(x) \right| < \varepsilon.$$

Thus,

$$\exists \delta > 0 \text{ with } 0 < |y-x| < \delta \Rightarrow \left| \frac{f(y)-f(x)}{y-x} - f'(x) \right| < \varepsilon.$$

In this classical formula, the parameters are f, ε, and x. They are all standard. Transfer can be applied and furnishes the same result for all $x \in J$ (standard subset of I) and all standard $\varepsilon > 0$. That gives precisely the conclusion we are looking for!

5.2 THEOREMS FOR DIFFERENTIABLE FUNCTIONS

5.2.1. Rolle's theorem

Let $I = [a, b] \subset \mathbb{R}$ be an interval with nonempty interior (i.e. $a < b$) and $f: I \to \mathbb{R}$ a continuous function with $f(a) = f(b)$. If f is differentiable at all interior points of I, then there is such an interior point $c \in]a, b[$ with $f'(c) = 0$.

5.2.2 Proof

If f is constant, there is nothing to prove (f' vanishes at all interior points). Otherwise, f has an extremum $f(c) \neq f(a)$ at an interior point $c \in]a, b[$. Necessarily $f'(c) = 0$. When f is standard, we see that f' vanishes at a standard interior point.

5.2.3 Corollary

If $f: I = [a, b] \to \mathbb{R}$ is continuous as before, differentiable at all interior points, then there is an interior point c of I with

$$f'(c) = \frac{f(b) - f(a)}{b - a}.$$

5.2.4 Proof

Rolle's theorem can be applied to the auxiliary function

$$F(x) = \begin{vmatrix} f(x) - f(a) & f(b) - f(a) \\ x - a & b - a \end{vmatrix} \quad \text{(two by two determinant)}$$

which vanishes at both ends of the interval (the first column is the zero column for $x = a$, whereas both columns are equal for $x = b$). Thus, there is an interior point c at which F' vanishes. But the derivative of the determinant is given by

$$F'(x) = \begin{vmatrix} f'(x) & f(b) - f(a) \\ 1 & b - a \end{vmatrix}$$

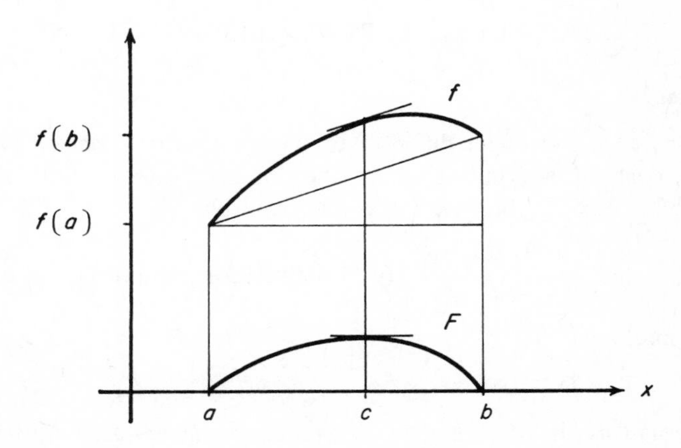

Fig. 5.1

(check it on the expansion of the determinant if you have any doubt!). But $F'(c)=0$ is precisely the announced equality.

Geometrically, F is obtained by subtracting a linear function to f in order to satisfy the assumptions of Rolle's theorem.

5.2.5 Application to monotonous functions

If f is a continuous real valued function defined on the interval $I=[a, b]$ (with nonempty interior as before), we see that f differentiable at all interior points with

$$m \leq f'(x) \leq M \quad \text{for all } x \in]a, b[$$

implies that

$$m \leq \frac{f(b)-f(a)}{b-a} \leq M$$

(indeed, the difference quotients are special values of the derivative by (5.2.3)).

This simple observation has the following consequences

$$f' \geq 0 \text{ on } I \Rightarrow f \quad \text{monotonous increasing,}$$
$$f' \leq 0 \text{ on } I \Rightarrow f \quad \text{monotonous decreasing,}$$

and in particular

$$f'=0 \text{ on } I \Rightarrow f \text{ constant.}$$

In other words, a function can have at most *one* primitive on an interval I, vanishing at a prescribed point

$$g=f'_1=f'_2 \Rightarrow (f_1-f_2)'=0 \Rightarrow f_1-f_2=\text{const.}$$

and thus $f_1=f_2$ as soon as these primitives take the same value at a given point.

5.3 STRICTLY DIFFERENTIABLE FUNCTIONS

5.3.1 Definition

Let us introduce a stronger differentiability condition. As before, we assume that f: $I=[a, b] \to \mathbb{R}$ is continuous on the compact interval with nonempty interior: $a<b$. Consider the function of two variables

$$g(x, y)=\frac{f(x)-f(y)}{x-y} \quad \text{(well defined for } x \neq y \text{ in } I\text{).}$$

More formally, let

$$D=\{(x, y) \in I \times I : x=y\}=\{(x, x) : x \in I\}$$

be the diagonal of the square $I \times I$, so that g is defined and continuous on $I \times I - D$.

Instead of defining notions of strict S-differentiability and strict differentiability having the same sense for standard functions and points, we shall content ourselves with defining this notion explicitly only for standard data (thus using the implicit definition obtained by transfer for the general case).

Let f and the interior point c be standard. Then we say that f is **strictly differentiable** at c if there exists a standard number m with

$$g(x, y) \simeq m \quad \text{for all } x \simeq c \simeq y \quad \text{and} \quad x \neq y \text{ (in } I).$$

Taking in particular $y=c$, we see that strict differentiability at c implies differentiability at c (with $m=f'(c)$). Intuitively, strict differentiability requires that slopes of secants going through two near points of $(c, f(c))$ on the graph of f should have a limit position (cf. Fig. 5.2. and Exercise (5.6.4)).

Strict differentiability is thus implicitly defined for all couples (f, c) by transfer. The interest of this notion comes from the following result.

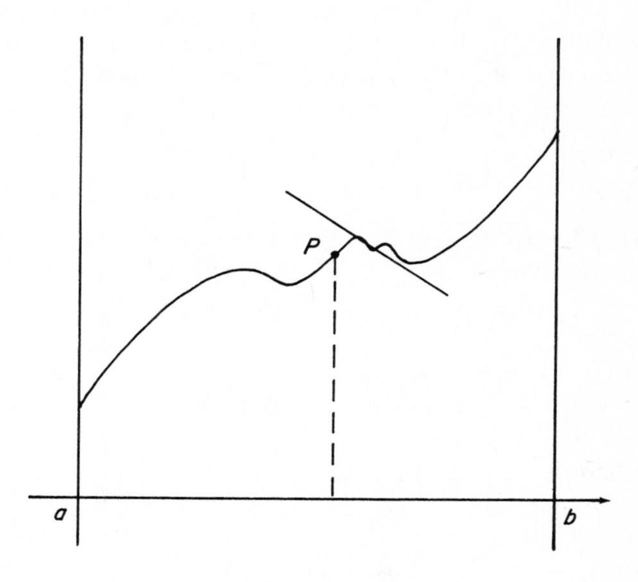

Fig. 5.2

5.3.2 Theorem

Let $f: I \to \mathbb{R}$ (or $\mathbb{C}$) be a standard function. The following two conditions are then equivalent.

(i) f is strictly differentiable at all interior points of I,
(ii) f is differentiable at all interior points of I and f' is a continuous function.

5.3.3 Proof

As we are assuming that f is standard, f' will automatically be standard.

Let us first prove (i)$\Rightarrow$(ii). Let us fix a standard interior point $a \in I$. By (4.1.2) it is enough to show that f' is S-continuous at a. Thus, we take $x \simeq a$ and we compare $f'(x)$ to $f'(a)$. By Proposition (5.1.9), we know that for $\delta > 0$ small enough

$$0 < |y-x| < \delta \Rightarrow \frac{f(y)-f(x)}{y-x} \simeq f'(x).$$

On the other hand, strict differentiability of f at a gives

$$\frac{f(y)-f(x)}{y-x} \simeq f'(a) \quad \text{if} \quad y \simeq x \simeq a,$$

and in particular also for all y sufficiently near to a. Comparison leads to $f'(x) \simeq f'(a)$.

Conversely, let us show (ii)$\Rightarrow$(i). To prove that f is strictly differentiable at all interior points of I, it is enough to prove it at all standard interior points. When $x \simeq a \simeq y$ (and $x \neq y$) Rolle's Theorem (5.2.1) furnishes a point z between x and y—hence *a fortiori* infinitely near a—with

$$\frac{f(y)-f(x)}{y-x} = f'(z).$$

Since we are assuming that f' is continuous, hence S-continuous at the standard point a, we have

$$f'(z) \simeq f'(a) = m$$

and thus

$$\frac{f(y)-f(x)}{y-x} \simeq m \quad \text{standard}$$

as was to be shown.

5.4 HIGHER DERIVATIVES

5.4.1 Limited expansions

There is no difficulty in introducing **higher derivatives**, defining f'' as the derivative of f' (whenever it exists), and $f^{(n)}$ as the derivative of $f^{(n-1)}$ (whenever it exists). If f is differentiable up to order n at a point a, it is possible to write a limited expansion

$$f(x) = f(a) + f'(a)\,(x-a) + \ldots + f^{(n)}\,(x)(x-a)^n/n! + r_{n,a}(x)$$

with a remainder $r_{n,a}(x)$ (precisely defined by the preceding equality . . .)

satisfying

$$r_{n,a}(x)/(x-a)^n \to 0 \quad \text{when } x \to a.$$

When (f, a) and n are standard, the preceding limited expansion is equivalent to

$$f(x) = f(a) + \ldots + f^{(n)}(a)(x-a)^n/n! + \varepsilon_{n,a}(x) \cdot (x-a)^n$$

where

$$\varepsilon_{n,a}(x) \simeq 0 \quad \text{when } x \simeq a.$$

There is no difficulty either in defining the notion of **infinitely differentiable** function. In this case, limited expansions can be written for all orders n, and in particular for illimited orders n.

5.4.2 Analytic functions

Let us still give a nonstandard characterization of analyticity (equivalent to the classical definition for standard data).

Let (f, a) be standard as before. Then we say that f is **analytic** at a if f is infinitely differentiable at a with

for n illimited, there is a standard $r > 0$ with

$$r_{n,a}(x) \simeq 0 \quad \text{for } |x-a| < r.$$

When this condition is satisfied, we can write a limited expansion of f to the order n illimited in the form

$$f(x) = p_{n,a}(x) + r_{n,a}(x)$$

with a polynomial $p_{n,a}$ of degree n (illimited) infinitely near f at all points of the standard neighbourhood $|x-a| < r$ of a. In this neighbourhood, f is the (strict) standard part of $p_{n,a}(x)$.

5.5 FINITE DIFFERENCES AND DERIVATIVES

5.5.1 Sequences

Let $(a_n)_{n \in \mathbb{N}}$ be a numerical sequence. We define the **difference operator** ∇ on such sequences by

∇a is the sequence with nth term equal to $a_{n+1} - a_n$.

For instance, if we take the sequence

$$n \mapsto \binom{n}{k} = \frac{n(n-1)\ldots(n-k+1)}{k!},$$

we get

$$\nabla \binom{n}{k} = \binom{n-1}{k-1}.$$

More generally, if $a=(a_n)$ is given by a linear combination of the preceding sequences, say

$$a_n = \sum_{0 \le k \le n} c_k \binom{n}{k},$$

we have

$$a_0 = c_0 \quad \left(\text{since} \binom{n}{k} = 0 \text{ if } k \ge 1\right)$$

as well as

$$\begin{aligned}(\nabla a)_n &= \sum_{0 \le k \le n} c_k \nabla \binom{n}{k} \\ &= \sum_{0 \le k \le n} c_k \binom{n-1}{k-1} = \sum_{1 \le k \le n} c_k \binom{n-1}{k-1}\end{aligned}$$

and thus

$$(\nabla a)_0 = c_1.$$

Repeating the argument, we find by induction

$$c_k = (\nabla^k a)_0.$$

In fact, the following result can easily be checked by induction on n.

5.5.2 Theorem

If $a=(a_n)_{n \in \mathbb{N}}$ is any sequence, then

$$\begin{aligned}a_n &= \sum_{0 \le k \le n} (\nabla^k a)_0 \binom{n}{k} = \sum_{0 \le k < \infty} (\nabla^k a)_0 \binom{n}{k} \\ &= \sum_{0 \le k < \infty} \frac{(\nabla^k a)_0}{k!} n(n-1) \ldots (n-k+1).\end{aligned}$$

5.5.3 Relation with derivatives

The preceding result should be compared with the Taylor series of an analytic function. In fact, the finite difference operator ∇ corresponds to a differential quotient with increment $h=1$. To understand this relation in a more precise form, let us consider a standard function $f\colon \mathbb{R} \to \mathbb{R}$ (or $\mathbb{C}$) which is N times differentiable at the origin. Then f has a limited expansion at the origin of the form

$$f(x)=\sum_{0\leq k\leq N}\frac{f^{(k)}(0)}{k!}x^k+x^N r_N(x)$$

where $r_N(x)\simeq 0$ if $x\simeq 0$.

Let us choose an infinitesimal $\varepsilon>0$ and consider the sequence $a_n=f(n\varepsilon)$. The finite difference expansion of this sequence can be written

$$a_n=\sum_{0\leq k<\infty}\frac{(\nabla^k a)_0}{k!}n(n-1)\ldots(n-k+1)$$

$$=\sum_{0\leq k<\infty}\frac{(\nabla^k a)_0}{k!\varepsilon^k}\varepsilon^k n^k(1-1/n)\ldots(1-(k-1)/n).$$

Comparison with the limited expansion at $x=n\varepsilon$

$$a_n=f(n\varepsilon)=\sum_{0\leq k\leq N}\frac{f^{(k)}(0)}{k!}\varepsilon^k n^k+(\varepsilon n)^N r_N(\varepsilon n)$$

can be made for n illimited, but such that εn is still infinitesimal (e.g. take for n the integral part of $1/\sqrt{\varepsilon}$). In this case, the expressions

$$(1-1/n),\ldots,(1-1/n)\cdot\ldots\cdot(1-k/n)$$

are infinitesimal for k limited.

We thus infer that if N is limited,

$$f^{(k)}(0)=\mathrm{st}\frac{(\nabla^k a)_0}{k}\quad\text{for}\quad k\leq N.$$

These formulas generalize the derivative formula

$$f'(0)=\mathrm{st}\frac{f(\varepsilon)-f(0)}{\varepsilon},$$

valid for a standard differentiable function f.

5.6 EXERCISES

5.6.1 Take any illimited integer $n\in\mathbb{N}$. Prove that

$$(1+x/n)^n\simeq e^x\quad\text{for every limited } x\in\mathbb{R}$$

(cf. Exercise 3.5.10).

5.6.2 Show that if a function f is S-differentiable at a point, then it is necessarily S-continuous at that point.

5.6.3 Discuss the differentiability of the functions $f_\varepsilon(x)=[x/\varepsilon]\varepsilon$ (defined for $x \geq 0$) when $\varepsilon > 0$ is infinitesimal.

5.6.4 Show that the function $f(x)=x^2 \sin(1/x)$ (with the convention $f(0)=0$) is differentiable, but not strictly differentiable at the origin $x=0$.

Still looking for a charmed (i.e. nonstandard) butterfly !

CHAPTER 6

INTEGRATION

6.1 METHOD

6.1.1 An example

To compute areas and volumes, the method of using infinitely thin slices is easily justified by nonstandard analysis. Let us show how the integral

$$\int_0^x t^2 \, dt$$

can be evaluated by this method.

For this purpose let us introduce a regular subdivision of the interval $[0, x]$: choose an integer $n > 0$ and introduce intermediate points $x_j = jh$ where $h = x/n$.

Adding the areas of the rectangles under the curve (as in the picture), we find

$$\sum_{0 \leq x_j < x} x_j^2 \cdot h = \sum_{0 \leq j < n} j^2 h^2 \cdot h = h^3 \sum_{j < n} j^2 .$$

But, it is well known (and easy to check by induction) that

$$1^2 + 2^2 + 3^2 + \ \ldots \ + n^2 = n(+1)(2n+1)/6,$$

and thus the preceding sum is

$$(x^3/n^3)(n-1)n(2n-1)/6 = (1 - 1/n)(1 - 1/2n)x^3/3.$$

In NSA, it is possible to choose n illimited (corresponding to h infinitesimal) and thus define the surface as the standard part of the corresponding sum (at least when x is standard). We indeed see that this standard part exists when x is standard and is $x^3/3$. We also observe that this result is independent from the illimited integer n chosen to find it.

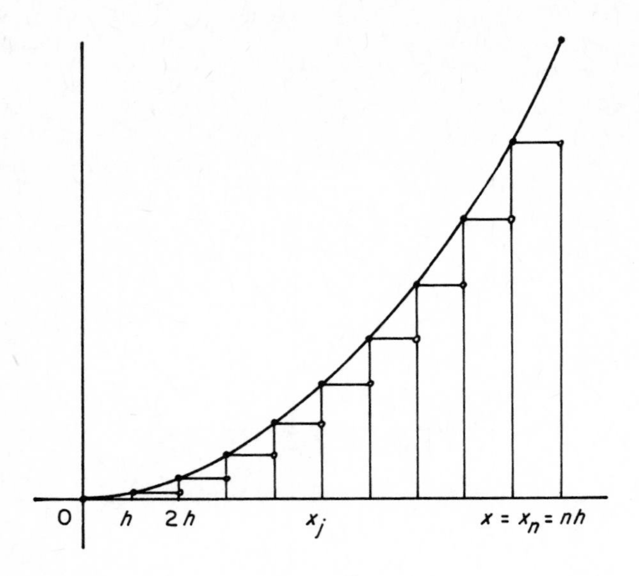

Fig. 6.1

6.1.2 Integration of powers

One can evaluate the integral of t^3 between 0 and x with the same method, provided we know the expression for the sum of cubes

$$1^3+2^3+3^3+\ldots+n^3.$$

Actually, sums of the kth powers were used for this purpose as early as the seventeenth century. Cavalieri, Fermat, Pascal, Torricelli, Wallis (and others . . .) all knew that

$$\int_0^x t^k \, \mathrm{d}t = x^{k+1}/(k+1).$$

The interested reader will be able to refer to the **Bernoulli formulas** for the sums

$$S_k(n) = 1^k + 2^k + 3^k + \ldots + n^k$$

to treat these definite integrals in a similar way. Explicitly,

$$S_1(n) = n(n+1)/2 = n^2/2 + \mathcal{O}(n),$$
$$S_2(n) = n(n+1)(2n+1)/6 = n^3/3 + \mathcal{O}(n^2),$$
$$S_3(n) = (S_1(n))^2 = n^2(n+1)^2/4 = n^4/4 + \mathcal{O}(n^3),$$
$$S_4(n) = n(n+1)(2n+1)(3n^2+3n-1)/30 = n^5/5 + \mathcal{O}(n^4), \text{ etc.}$$

The crucial observation is

$$S_k(n) = n^{k-1}/(k+1) + \mathcal{O}(n^k) \quad \text{for all integers } k \in \mathbb{N},$$

since it allows one to estimate the standard part of the finite (but illimited) sum approximating the integral of the kth power precisely as we did for the second power.

6.2 DEFINITE INTEGRAL

6.2.1 Definition of the integral

Let $a < b$ and $f: I = [a, b] \to \mathbb{R}$ (or $\mathbb{C}$) be a bounded function. Fix an infinitesimal $h > 0$. We wish to define the **integral of** f on I by

$$\int_I f = \int_a^b f(x)\,\mathrm{d}x = st \sum_{a \le jh < b} f(jh) \cdot h.$$

To be able to do this, and in particular to be able to speak of this standard part, it is necessary to check that the sum is limited. But each term in the sum is majorized by $h \cdot \sup|f|$ and the number of terms of the sum is also majorized by $(b-a)/h + 1$. Thus,

$$\left|\sum f(jh)h\right| \le \sum \left|f(jh)h\right| \le \sup|f| \cdot h \cdot \left(\frac{b-a}{h} + 1\right)$$

$$\le (b-a+h)\sup|f|.$$

In particular, the sum will be limited if f, a and b are standard.

Let us thus *define* the **definite integral** as the standard map

$$\Theta: (a, b\, f) \mapsto \int_a^b f(x)\,\mathrm{d}x$$

which takes the (standard) values

$$\Theta(a, b, f) = \mathrm{st} \sum_{a \le jh < b} f(jh) \cdot h$$

on the standard triples $a \le b \in \mathbb{R}$, f bounded $I = [a, b] \to \mathbb{R}$ of $\mathbb{C}$.

6.2.2 Properties of the integral

We observe immediately that if $a < b < c$ and f is bounded on $[a, c]$, then

$$\int_a^c f(x)\,\mathrm{d}x = \int_a^b f(x)\,\mathrm{d}x + \int_b^c f(x)\,\mathrm{d}x.$$

Indeed, it is sufficient to check this formula when all parameters are standard (by transfer). In this particular case, it simply follows from the obvious identity

$$\sum_{a\leq jh<c} f(jh)\cdot h=\sum_{a\leq jh<b} f(jh)\cdot h+\sum_{b\leq jh<c} f(jh)\cdot h,$$

and then taking the standard part of both sides (the standard part of a sum is the sum of the standard parts, cf. (3.2.6)).

We see here the interest of taking a fixed subdivision

$$\mathbb{Z}h=\{jh: j\in\mathbb{Z}\subset\mathbb{R} \text{ (or of any interval } I=[a, b])\}.$$

It would be impossible to prove the preceding identity simply with regular subdivisions

$$\begin{aligned} x_i&=a+i(c-a)/n \quad \text{of } [a, c],\\ \xi_j&=a+j(b-a)/\nu \quad \text{of } [a, b],\\ \eta_k&=b+k(c-b)/\mu \quad \text{of } [b, c] \end{aligned}$$

(n, ν and μ being illimited integers), simply because we have made insufficient assumptions on f. . . .

6.2.3 A further observation

Coming back to the definition of the definite integral, it is also obvious that for $f\leq g$

$$\sum_{a\leq jh<b} f(jh)\cdot h\leq\sum_{a\leq jh<b} g(jh)\cdot h,$$

and hence

$$\int_a^b f(x)\,\mathrm{d}x\leq\int_a^b g(x)\,\mathrm{d}x$$

(first for f, g, a and b standard . . .). In particular,

$$m\leq f\leq M \text{ on } [a, b]$$

implies

$$m(b-a)\leq\int_a^b f(x)\,\mathrm{d}x\leq M(b-a).$$

Taking $m=-\sup|f|$, $M=\sup|f|$, we have obtained

$$\left|\int_a^b f(x)\,\mathrm{d}x\right|\leq(b-a)\sup|f|$$

(essentially, this inequality has already been obtained in the preceding Section (6.1.1)).

The following consequence results immediately from this.

If $f(x) \simeq 0$ for $a \leq x \leq b$ with a and b limited, then also

$$\int_a^b f(x)\,\mathrm{d}x \simeq 0.$$

The context in which we have defined the definite integral is of course a bit artificial. For example, it is obvious that the integral only depends (at least when all data are standard) on the values of f on the fixed subdivision consisting of integral multiples of the infinitesimal $h > 0$. This integral depends thus on the choice of h. The next result shows that for *continuous* functions, the definition is relevant!

6.2.4 Theorem

As previously defined, the integral of a continuous function on a compact interval is independent from the particular choice of infinitesimal used to compute it.

6.2.5 Proof

We assume f standard, continuous of $I = [a, b]$ compact (and real valued). For standard points $x_1 < x_2$ in I, we have

$$(x_2 - x_1) \cdot \min f \leq \int_{x_1}^{x_2} f(x)\,\mathrm{d}x \leq (x_2 - x_1) \cdot \max f$$

where the min and max of f can be taken on the interval $[x_1, x_2]$. By transfer, these inequalities remain true for all f, x_1, x_2 (provided f is continuous on $[x_1, x_2]$. Transfer is indeed legitimate since the integral could be written in the form

$$\Theta(f, x_1, x_2)$$

with an extra-parameter Θ representing a standard function. For a fixed $x_0 \in [x_1, x_2]$ we infer that

$$\min f - f(x_0) \leq \frac{1}{x_2 - x_1} \int_{x_1}^{x_2} f(x)\,\mathrm{d}x - f(x_0) \leq \max f - f(x_0).$$

Take now $x_1 \simeq x_0 \simeq x_2$, in which case the conclusion is that

$$x \mapsto F(x) = \int_a^x f(t)\,\mathrm{d}t$$

is strictly differentiable at x_0 (standard first) with derivative equal to $f(x_0)$. Hence F is a primitive of f. Similarly, if $G(x)$ denotes the integral obtained with another choice of infinitesimal k in place of h, G will also be a primitive of f. But we have seen that $F' = G'$ $(= f)$ implies $F - G$ is constant in the interval $[a, b]$ (cf. (5.2.5)). Since both primitives vanish at a, they must coincide, thereby proving the independence of the integral from the choice of infinitesimal chosen to compute it.

6.3 STRICT STANDARD PART OF A FUNCTION

6.3.1 A formal computation

Let us motivate a definition by the consideration of a formal computation (which could have been made by Euler. . .).

Assume that to differentiate the exponential, we start with the observation

$$e^x = \mathrm{st}\,(1 + x/n)^n$$

valid for x standard and n illimited (cf. Exercise (5.5.1)). The derivative of the polynomial $(1 + x/n)^n$ is

$$n(1 + x/n)^{n-1}/n = (1 + x/n)^n(1 + x/n)^{-1}$$

$$\simeq (1 + x/n)^n \simeq e^x \quad \text{if } x \text{ is limited and } n \text{ illimited.}$$

Thus we guess that the derivative of e^x is the function e^x itself!

This procedure finds a natural justification in nonstandard analysis.

6.3.2 Justification

In place of the exponential, take any standard function

$$f: I \to \mathbb{R} \text{ where the interval } I \subset \mathbb{R} \text{ is (standard) open } \neq \varnothing.$$

Suppose that a polynomial p with $p(x) \simeq f(x)$ (all $x \in I$) has been chosen, and there exists a standard function g with $p'(x) \simeq g(x)$ (all $x \in I$). In the preceding example, the standard function g was again the exponential. Then we expect that f is differentiable with $f' = g$. This is precisely the content of next result.

6.3.3. Theorem

Let f be a standard function (defined on some standard open interval of $\mathbb{R}$) and p a continuously differentiable function with $f(x) \simeq p(x)$ at all x (in the considered interval). Then, if there is a standard function g with $g(x) \simeq p'(x)$ at all x, then f is continuously differentiable with the derivative $f' = g$.

6.3.4 Proof

Take a standard point $a \in I$ (definition interval of f). For x standard first, we can write

$$f(x)-f(a) \simeq p(x)-p(a) \simeq \int_a^x p'(t)\, dt \simeq \int_a^x g(t)\, dt.$$

The extreme terms of the preceding chain of 'near equalities' are standard (f, g, x and a are standard) hence equal

$$f(x)-f(a) = \int_a^x g(t)\, dt.$$

By transfer, this equality will persist for all x (standard or not). This proves that f is strictly differentiable (Section 5.3) with a derivative equal to g.

6.3.5 Definition

We shall say that a function f admits a strict standard part when there exists a standard function g with $f(x) \simeq g(x)$ at *all points* (in the domain of definition of f).

When $f(x)$ is limited for all standard x, we have already defined the standard part f^* of f (cf. (3.2.9)). But in general, this function will not remain infinitely close to f at all points. If it does, this standard part f^* is strict.

6.3.6 Reformulation of the theorem

In the situation of (6.3.3), the continuously differentiable function p was assumed to have a strict standard part f and p' a strict standard part g. Then f is continuously differentiable with $f' = g$. Thus, formal computations are justified provided the standard parts are strict.

6.4 EXERCISES

6.4.1 Compute the definite integral $\int_0^1 e^x\, dx$ by taking a regular subdivision 0, h, $2h, \ldots, nh = 1$ of the interval [0, 1]. Use the same method for the integral

$$\int_0^1 xe^x\, dx.$$

6.4.2 Compute the integral

$$\int_0^\pi \log(1-2a\cos x + a^2)\, dx \quad \text{for } 0 < a \neq 1,$$

using an infinitesimal h of the form π/n (with n illimited).

6.4.3 Justify the following deduction (due to Leibniz!)

We start with the identity

$$1-x+x^2-\ \ldots\ =1/(1+x)\quad \text{(valid for } |x|<1)$$

which we integrate term by term getting

$$x-x^2/2+x^3/3-\ \ldots\ =\log(1+x)$$

(substituting $x=0$ indeed shows that the integration constant vanishes since $\log 1=0$). Taking $x=1$ (!?) gives then

$$1-1/2+1/3-\ \ldots\ =\log 2.$$

SECOND PART

CHAPTER 7

INVARIANT MEANS

7.1 DEFINING PROPERTIES

7.1.1 Classical properties required

When we try to define a notion of natural density for parts of integers, it is natural to consider maps

$$m: \mathscr{P}(\mathbb{Z}) \rightarrow [0, 1]$$

on the set of all parts of $\mathbb{Z}$ and having the following properties

(a) *normalization*: $m(\mathbb{Z}) = 1$,

(b) *additivity*:

$$m(A \cup B) = m(A) + m(B) \quad \text{if } A \text{ and } B \text{ are disjoint,}$$

(c) *translation invariance*:

$$A_t = \{a + 1 : a \in A] \Rightarrow m(A_t) = m(A).$$

Such a map is called an **invariant mean** (on the integers $\mathbb{Z}$).

7.1.2 First consequences of postulated properties

By induction, the additivity holds for finite families of disjoint part of $\mathbb{Z}$: a **measure** would even be required to have the additivity property for *countable* families of disjoint parts. But in our context, an invariant measure would give same 'volume' to all points (by translation invariance) and countable additivity would imply $m(\mathbb{Z}) = \infty$ if $m(\{0\}) > 0$. Thus, normalization could not be realized.

Observe also that if $A \subset C$, we can always write C as disjoint union $C = A + B$

(with B = complement of A in C!) and in particular

$$m(A) \leq m(A) + m(B) = m(A \cup B) = m(C),$$

thereby proving that m is a monotonous (increasing) map.

Assume for an instant that an invariant mean exists on $\mathbb{Z}$. Consider the parts

$$A = N\mathbb{Z}, \qquad A_t = 1 + N\mathbb{Z}, \ldots$$

which are obtained from A by successive unit translations. There are N such disjoint parts which give a partition of the whole set of integers. Hence, the postulated properties of m immediately imply

$$m(N\mathbb{Z}) = 1/N.$$

For example, the measure of the even integers is 1/2, and so is the measure of the odd numbers.

Still assume that an invariant mean m exists. Then for each finite part $A \subset \mathbb{Z}$, say $A \subset [-n, n[$, all translated parts obtained by adding $2n$, $4n$, ..., $2kn$ respectively, are disjoint so that additivity implies

$$km(A) = m((A + 2n) \cup \ldots \cup (A + 2kn)) \leq m(\mathbb{Z}) = 1.$$

Thus, $m(A) \leq 1/k (k \geq 1)$ and $m(A) = 0$ for every finite part $A \subset \mathbb{Z}$. In particular, an invariant mean is never a measure.

7.2 EXISTENCE OF INVARIANT MEANS

7.2.1 A nonstandard construction

It is not completely obvious that an invariant mean exists on the set $\mathbb{Z}$ of integers. Here is one way of constructing such an object with NSA. Put first,

$$m_n(A) = \frac{\text{Card}(A \cap [-n, n])}{2n+1} \quad \text{for integers} \quad n \in \mathbb{N}.$$

As $2n + 1 = \text{Card}[-n, n]$, it is clear that requirement (a) of (7.1.1) is satisfied; (b) is also satisfied. Moreover, the intersections of A and A_t with $[-n, n]$ have nearly the same number of elements: precisely

$$\text{Card}(A_t \cap [-n, n]) - \text{Card}(A \cap [-n, n]) = -1, \qquad 0 \text{ or } +1.$$

Thus,

$$|m_n(A_t) - m_n(A)| \leq 1/(2n+1)$$

and this suggests that the condition (c) of (7.1.1) will be satisfied if we take n illimited and define m by taking the standard part of m_n. Choose an illimited integer $n \in \mathbb{N}$ and *define*

$$m(A) = \text{st}\, m_n(A) \quad \text{for a standard part } A \subset \mathbb{Z}.$$

There is one and only one standard map m: $\mathscr{P}(\mathbb{Z}) \to [0, 1]$ taking the previously defined values on standard parts A of $\mathbb{Z}$. For such a map, all properties of invariant means are satisfied. For example, if the part A is standard, A_t is also standard and

$$m(A_t) - m(A) \text{ standard and infinitesimal, vanishes by (3.2.3).}$$

This proves translation invariance of m (on standard parts first, hence for all parts by transfer).

7.2.2 An example

Since the mean of a finite part is zero, we can always delete a finite number of points of a part A without changing the value of $m(A)$. In particular, all parts $A_n = A \cap \{x \in \mathbb{N} : |x| > n\}$ have the same invariant mean. Thus, an invariant mean has a limit meaning and from a classical point of view, we might be tempted to consider

$$\lim_{n\to\infty} m_n(A)\,(=m(A)?).$$

Unfortunately, such a limit does not exist for all parts $A \subset \mathbb{Z}$ as the following example shows. Let

$$A = \{m \in \mathbb{Z} \colon \exists k \in \mathbb{N} \quad \text{with} \quad 3^{2k} < |m| \leq 3^{2k+1}\}.$$

This set is sketched in the following picture. Taking for n an odd power of 3, say $n = 3^{2k+1}$, we find for this particular part $A, m_n(A) \geq 2/3$. Taking for n an even power of 3, say $n = 3^{2k+2}$, we find on the contrary $m_n(A) \leq 1/3$!

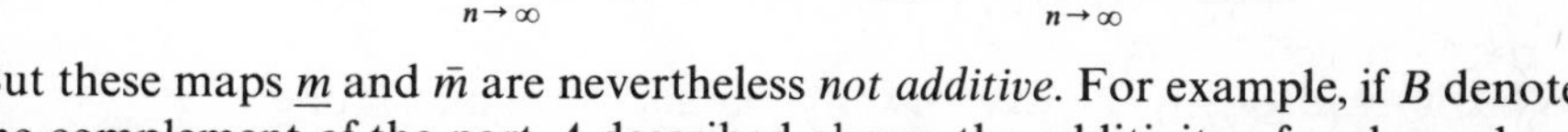

Fig. 7.1

A second idea would be to consider

$$\bar{m}(A) = \limsup_{n\to\infty} m_n(A) \quad \text{and} \quad \underline{m}(A) = \liminf_{n\to\infty} m_n(A).$$

But these maps $\underline{m}$ and $\bar{m}$ are nevertheless *not additive*. For example, if B denotes the complement of the part A described above, the additivity of each m_n shows that for n equal to an even power of 3, $m_n(B) \geq 2/3$. Thus

$$\bar{m}(A) = \limsup_{n\to\infty} m_n(A) \geq 2/3, \qquad \bar{m}(B) = \limsup_{n\to\infty} m_n(B) \geq 2/3$$

in spite of the fact that A and B make up a disjoint decomposition of $\mathbb{Z}$ in this case.

7.2.3 Construction with ultrafilters

A classical procedure would consist in choosing a *nonprincipal ultrafilter* $\mathscr{U}$ on $\mathbb{N}$ and defining

$$m(A) = \lim_{\mathscr{U}} m_n(A).$$

This gives a (nonconstructive) way of defining invariant means which is, in fact, completely analogous to the NSA procedure. It is easy to imagine that the invariant mean $m = m_{\mathscr{U}}$ defined with the ultrafilter $\mathscr{U}$ indeed depends on this choice (we shall see below that the invariant mean defined in NSA by $m = \mathrm{st}(m_n)$ also depends on the choice of the illimited integer n).

7.3 A RESULT

7.3.1 Proposition

Let n be an illimited integer and m the invariant mean on $\mathbb{Z}$ defined by $m(A) = \mathrm{st}\, m_n(A)$ for standard parts $A \subset \mathbb{Z}$. Then, with the notations of (7.2.2),

$$\underline{m}(A) \leq m(A) \leq \bar{m}(A).$$

7.3.2 Proof

It is enough to establish the inequalities for A standard. Then, for each standard $N \in \mathbb{N}$, we have $n > N$ and hence

$$\sup_{k \geq N} m_k(A) \geq m_n(A).$$

Taking standard parts (cf. 3.2.6), we infer

$$\sup_{k \geq N} m_k(A) \geq m(A)$$

since the term on the left side of the preceding inequality is standard (indeed, the parameters A and N are standard). Thus, we must also have

$$\sup_{k \geq N} m_k(A) \geq m(A)$$

for all integers $N \in \mathbb{N}$ by transfer (the standard map m acting as standard value of a parameter in the formula). Then, going to the limit $N \to \infty$, we conclude $\bar{m}(A) \geq m(A)$ as expected.

The other inequality concerning m is obtained symmetrically.

7.4 MORE COMMENTS

7.4.1 Back to Example (7.2.2)

Going back to the Example 7.2.2 above, we see that if we take the illimited integer n to be an even power of 3, say $n = 3^{2k}$ with an illimited integer k, then we get

$$\mathrm{Card}(A \cap [0, n]) = 2 \cdot 3^{2k-2} + \ldots + 2 \cdot 3^2 + 2 = 2 \cdot \frac{3^{2k} - 1}{9 - 1} = \frac{1}{4}(3^{2k} - 1)$$

so that finally

$$\text{st}\, m_n(A) = 1/4.$$

If, on the contrary, we choose n to be an illimited odd power of 3, we shall obtain

$$\text{st}\, m_n(A) = 3/4.$$

Since A is a standard part in this example, we see that $m(A) = 1/4$ for the first invariant mean obtained, whereas $m(A) = 3/4$ for the second invariant mean. This exhibits a dependence on the invariant mean according to the choice of the illimited integer chosen to define it.

On the other hand, the property

$$m(N\mathbb{Z}) = 1/N = m(\mathbb{Z})/N$$

would suggest that perhaps $m(NA) = m(A)/N$ for all parts A and all integers N (NA denotes the set consisting of all Na for $a \in A$). However, the Exercise 7.5.3 shows that this is not the case.

7.4.2 Invariant means for bounded functions

An invariant mean m on $\mathbb{Z}$ can be used to measure means of bounded functions on $\mathbb{Z}$. Denote by $l^\infty = l^\infty(\mathbb{Z})$ the real (or complex) vector space of bounded functions, (i.e. bounded sequences)

$$f\colon \mathbb{Z} \to \mathbb{R}\,(\text{resp. } \mathbb{C}).$$

For such a function f, and two given $a < b \in \mathbb{Z}$, introduce the notation

$$m(a \le f < b) = m\{x \in \mathbb{Z} : a \le f(x) < b\}.$$

Now, take an infinitesimal $h > 0$. For each standard $f \in l^\infty$, the sum

$$\sum_{k \in \mathbb{Z}} m(kh \le f < (k+1)h) \cdot kh$$

is limited (bounded in absolute value by $\| f \| + h = \sup |f| + h$). It is thus possible to define

$$m(f) = \text{st} \sum_{k \in \mathbb{Z}} m(kh \le f < (k+1) \cdot kh$$

when f is a standard bounded function. There is only one standard map (still denoted m)

$$m : l^\infty \to \mathbb{R}$$

taking the previously prescribed values on standard elements. We have

$$|m(f)| \le \|f\| = \sup |f| \quad \text{and} \quad m(f_a) = m(f)$$

if f_a denotes a translate of f: $f_a(x) = f(x - a)$. Thus, an invariant mean defines a continuous linear form of l^∞, namely, an element of the Banach space $(l^\infty)'$ dual of l^∞ (the linear forms thus obtained have the further translation invariance property).

7.4.3 Theorem

If the invariant mean $m = \text{st}\, m_n$ is defined as above with the choice of an illimited integer n, then, for a standard $f \in l^\infty$

$$m(f) = \text{st} \frac{1}{2n+1} \sum_{|k| \leq n} f(k).$$

7.4.4 Proof

Let $I = [-n, n]$ be the basic interval used for the definition of m_n (and of m). For any integer k, define

$$A_k = \{kh \leq f < (k+1)h\} \cap I.$$

We have

$$0 \leq f(a) - kh < h \quad \text{for} \quad a \in A_k$$

and by summation over all $a \in A_k$

$$0 \leq \sum_{A_k} (f(a) - kh) < \text{Card}(A_k) \cdot h.$$

Hence

$$0 \leq (2n+1)^{-1} \sum_{A_k} f(a) - (2n+1)^{-1} \sum_{A_k} kh < m_n(A_k) \cdot h,$$

$$0 \leq (2n+1)^{-1} \sum_{A_k} f(a) - (2n+1)^{-1} \text{Card}(A_k) kh < m_n(A_k) \cdot h.$$

Summing over k, we find

$$0 \leq (2n+1)^{-1} \sum_{I} f(a) - m_n(f) < h,$$

i.e.

$$(2n+1)^{-1} \sum_{I} f(a) \simeq m_n(f).$$

The announced result follows from this and the fact that

$$m(f) = \text{st}\, m_n(f) \quad \text{when} \quad f \text{ is standard.}$$

Alternatively, the reader can prove the theorem along the following lines. Check it first when f is standard and constant. Check it then when f standard only takes a finite number of values. Finally, the general case follows by uniform approximation.

7.4.5 Comparison with the Haar measure of $\hat{\mathbb{Z}}$

On the compact totally discontinuous group

$$\hat{\mathbb{Z}} = \prod_{p \in P} \mathbb{Z}_p = \lim_{\longleftarrow} (\mathbb{Z}/n\mathbb{Z})$$

(P denotes the set of primes and the projective limit is taken with respect to the divisibility relation and the canonical projection homomorphisms $\mathbb{Z}/nm\mathbb{Z} \to \mathbb{Z}/n\mathbb{Z}$) there is a normalized *Haar measure* μ. This measure satisfies

$$\mu(N\hat{\mathbb{Z}}) = \mu(\Pi p^{v_p(N)} \mathbb{Z}_p) = \Pi p^{-v_p(N)} \cdot \mu(\hat{\mathbb{Z}}) = 1/N.$$

Hence

$$\mu(N\hat{\mathbb{Z}}) = m(N\hat{\mathbb{Z}}) \quad \text{for all positive } N \in \mathbb{N}.$$

These equalities suggest that we introduce the canonical injection of completion $i: \mathbb{Z} \to \hat{\mathbb{Z}}$, and that we compare the invariant mean $m(A)$ with the Haar measure $\mu(\hat{A})$ of the closure $\hat{A}$ (or completion) of $i(A)$ in $\hat{\mathbb{Z}}$. One can then prove

$$1 - \mu(\hat{B}) \le m(A) \le \mu(\hat{A}) \quad \text{if} \quad B = A' = \hat{\mathbb{Z}} - A.$$

Thus, the value of $m(A)$ is well determined (independent from the choice of illimited n chosen to define m) on all parts A with

$$\hat{A} \cap \hat{B} \text{ negligible for the Haar measure of } \hat{\mathbb{Z}}.$$

Restricting one's considerations to these regular parts A of $\hat{\mathbb{Z}}$, we would have $m(NA) = m(A)/N$, and m would be countably additive.

7.5 EXERCISES

7.5.1 Let m be an invariant mean on $\mathbb{Z}$ defined by the choice of an illimited integer n. Discuss the following apparent paradox. If $A = [-n, n]$, we have $m_n(A) = 1$ whence $\operatorname{st} m_n(A) = 1$. However, we said at the end of 7.1.2 that $m(A) = 0$ for all finite $A \subset \mathbb{Z}$ (is this only true for standard finite subsets A?).

7.5.2 We keep the notations of Section 7.2 concerning the nearly invariant means m_n (n illimited integer). Prove that

$$m_n(\mathbb{N}) \simeq 1/2,$$

and hence $m(\mathbb{N}) = 1/2$ independently from the choice of n. Can you prove that $m(\mathbb{N}) = 1/2$ for *any* invariant mean m on $\mathbb{Z}$?

7.5.3 Compute the invariant mean $m(3A)$ as a function of $m(A)$ for the part A given in 7.2.2 Taking $m = \operatorname{st} m_n$ with an illimited power of 3 for n, show that $m(3A) \neq m(A)/3$.

7.5.4 Define an invariant mean m on the group $\mathbb{Z}^2$ by a method similar to that used in 7.2. Is the following deduction correct? For every integer $d>0$, we denote by A_d the part of $\mathbb{Z}^2$ consisting of all couples (a, b) having greatest common divisor d

$$A_d=\{(a, b)\in\mathbb{Z}^2 : gcd(a, b)=d\}.$$

These are disjoint subsets with union equal to $\mathbb{Z}^2-(0, 0)$. Hence

$$\sum_{d\geq 1} m(A_d)=m(\cup A_d)=m(\mathbb{Z}^2-(0,0))=m(\mathbb{Z}^2)=1.$$

Moreover, since

$$gcd(a, b)=d\Rightarrow gcd(a/d, b/d)=1,$$

we have $A_d=dA_1$, $m(A_d)=m(A_1)/d^2$. This gives

$$m(A_1)\sum_{d\geq 1} 1/d^2=1.$$

Finally, since we know that the sum of all inverses of squares is $\zeta(2)=\pi^2/6$, we infer

$$m(A_1)=6/\pi^2=0.6079271018\ldots$$

(In other words, the probability of selecting a relatively prime couple among all couples of integers is $6/\pi^2=0.6079271018\ldots$.)

CHAPTER 8

APPROXIMATION OF FUNCTIONS

8.1 DIRAC FUNCTIONS

8.1.1 Definition

A **Dirac function** $D\colon \mathbb{R} \to \mathbb{R}$ is a function possessing the following three basic properties

- D is *positive* and *integrable*,
- *normalization*: $\int D(x)\,\mathrm{d}x = 1$,
- *localization*: there exists an infinitesimal $\delta > 0$ with

$$\int_{|x|<\delta} D(x)\mathrm{d}x \simeq 1.$$

Obviously, these properties imply

$$\int_{|x|\geq\delta} D(x)\,\mathrm{d}x \simeq 0$$

so that the whole 'mass' of D is concentrated in the infinitesimal neighbourhood $|x| < \delta$ of 0.

8.1.2 Proposition

Let D be a Dirac function and f a standard continuous function. Then

$$\mathrm{st} \int f(x)D(x)\,\mathrm{d}x = f(0).$$

8.1.3 Proof

Let $\delta>0$ be an infinitesimal for which the third defining property of Dirac function is satisfied (for D). Since f is also S-continuous at the origin, we have

$$\int_{|x|<\delta} (f-f(0))D \, \mathrm{d}x \simeq 0.$$

We also have

$$\int_{|x|\geq\delta} (f-f(0))D \, \mathrm{d}x \simeq 0$$

since

$|f-f(0)| \leq 2\|f\|$ limited (because standard),

$$\int_{|x|\geq\delta} D(x) \, \mathrm{d}x \simeq 0.$$

Hence $0 \simeq \int (f-f(0))D \, \mathrm{d}x = \int fD \, \mathrm{d}x - f(0) \int D \, \mathrm{d}x$.

The proof is completed and the proposition shows that

$$f \mapsto \mathrm{st} \int fD \, \mathrm{d}x$$

implicitly defines the Dirac evaluation linear form $f \mapsto f(0)$ on the space of continuous functions. In fact, the preceding proof only requires

f integrable and bounded,
$\|f\|$ limited,
f S-continuous at the origin.

8.1.4 Proposition

Let f be standard, positive, integrable with $\int f \, \mathrm{d}x = 1$. Then $D(x) = nf(nx)$ is a Dirac function if $n \in \mathbb{N}$ is illimited.

8.1.5 Proof

By assumption, the function $R \mapsto \int_{|x|\geq R} f \, \mathrm{d}x$ is standard and tends to zero. Consequently,

$$\int_{|x|\geq R} f \, \mathrm{d}x \simeq 0 \quad \text{as soon as } R \text{ is illimited.}$$

Choose now $\delta = 1/\sqrt{n}$ (where $n \in \mathbb{N}$ is illimited). We have

$$\int_{|x| \geq \delta} nf(nx)\,\mathrm{d}x = \int_{|\xi| = |nx| \geq n\delta = \sqrt{n}} f(\xi)\,\mathrm{d}\xi \simeq 0.$$

8.1.6 Examples

As the preceding proposition shows, the following functions are Dirac functions when the integer n is illimited

$$D(x) = \begin{cases} n & \text{if } |x| \leq 1/(2n) \\ 0 & \text{otherwise,} \end{cases}$$

$$D(x) = \frac{n}{\pi(1 + n^2x^2)},$$

$$D(x) = n \cdot \exp(-\pi n^2 x^2).$$

8.1.7 Remark

A Dirac function is never a standard function. Indeed, for any standard integrable function f, the sequence

$$n \mapsto \int_{|x| \leq 1/n} |f(x)|\,\mathrm{d}x$$

is standard and tends to 0. Hence

$$\int_{|x| \leq 1/n} |f(x)|\,\mathrm{d}x \simeq 0 \quad \text{for all illimited } n.$$

The localization property is thus never attained for standard functions.

8.2 PERIODIC FUNCTIONS AND TRIGONOMETRIC POLYNOMIALS

8.2.1 Review of the Fourier theory

Let $f\colon \mathbb{R} \to \mathbb{R}$ (or $\mathbb{C}$) be a 1-periodic function, namely

$$f(x+1) = f(x) \quad \text{for all } x \in \mathbb{R}.$$

Examples of such functions abound: the basic exponentials

$$e_n(x) = e^{2\pi i n x} = e(x)^n \text{ (where } e(x) = e^{2\pi i x})$$

are such functions (for $n \in \mathbb{Z}$). If the function f is integrable, we can define the

Fourier coefficients of f by the usual formulas

$$c_n = c_n(f) = \int_0^1 f(x)\, e^{-2\pi i n x}\, \mathrm{d}x = (e_n | f),$$

with the scalar product

$$(g | f) = \int_0^1 \overline{g(x)} \cdot f(x)\, \mathrm{d}x.$$

Let us still denote by

$$S_N = S_N(f) = \sum_{|n| \leq N} c_n \cdot e_n$$

the symmetric partial sums of the Fourier series $\Sigma_{n \in \mathbb{Z}}\, c_n \cdot e_n$. The *Fejer averages* of f are defined by

$$\begin{aligned} A_N = A_N(f) &= (S_0 + S_1 + \ldots + S_{N-1})/N \\ &= \sum_{|n| \leq N} \frac{N - |n|}{N}\, c_n \cdot e_n. \end{aligned}$$

These expressions are trigonometric polynomials and we intend to prove the following basic convergence result.

8.2.2 Fejer's Theorem

Let f be continuous and 1-periodic over the reals. Then f is a uniform limit of trigonometric polynomials. More precisely, the Fejer averages $A_N(f)$ converge uniformly to the function f.

8.2.3 Proof

As usual, it is enough to prove the statement when f is a standard function. Then, $A_N(f)$ is a standard sequence and it is enough (by 3.4.7) to prove

$$A_N(f)(x) \simeq f(x) \quad \text{for all } x \in \mathbb{R} \quad \text{if } N \text{ is illimited.}$$

First we estimate the symmetric sums $S_N(f)$ using the defining expressions for the Fourier coefficients c_n (to simplify notations, we adopt the convention that all integrals $\int$ have to be computed on any interval of unit length). Thus

$$\begin{aligned} S_N(f)(x) &= \sum_{|n| \leq N} e_n(x) \cdot \int e_n(-y) f(y)\, \mathrm{d}y \\ &= \int f(y) \sum_{|n| \leq N} e_n(x - y)\, \mathrm{d}y = \int f(y) K_N(x - y)\, \mathrm{d}y. \end{aligned}$$

Let us compute

$$\begin{aligned} K_N(t) &= \sum_{|n|\le N} e_n(t) = \sum_{|n|\le N} e(t)^n \\ &= e(t)^{-N}(1+e(t)+\ \dots\ +e(t)^{2N}) \\ &= e(t)^{-N}\frac{e(t)^{2N+1}-1}{e(t)-1} = \frac{e(t)^{N+1/2}-e(t)^{-N-1/2}}{e(t)^{1/2}-e(t)^{-1/2}} \\ &= \sin(N+1/2)2\pi t/\sin \pi t. \end{aligned}$$

The formulas for the averages are similar. Explicitly,

$$A_N(f)(x) = \int f(y)\frac{1}{N}\sum_{0\le n<N} K_n(x-y)\,dy$$

and we still have to evaluate

$$\frac{1}{N}\sum_{0\le n<N} \sin(n+1/2)2\pi t.$$

For this purpose, we observe that this sum is the imaginary part of

$$\begin{aligned} &e(t/2)(1+e(t)+\ \dots\ +e(t)^{N+1}) \\ &= e(t/2)\frac{e(t)^N-1}{e(t)-1} = \frac{e(t)^N-1}{2i}\,\frac{1}{\sin \pi t}. \end{aligned}$$

But now,

$$\operatorname{Im}(e(t)^N-1)/2i = \operatorname{Re}(1-e(t)^N)/2 = (1-\cos 2N\pi t)/2 = \sin^2 N\pi t$$

and the sum of the geometric series is

$$\sin^2 N\pi t/\sin \pi t$$

whereas

$$K_0(t)+\ \dots\ +K_{N-1}(t) = \sin^2 N\pi t/\sin^2 \pi t.$$

Altogether, we have found

$$A_N(f)(x) = \int f(y)D_N(x-y)\,dy$$

where

$$D_N(t) = \frac{1}{N}\sin^2 N\pi t/\sin^2 \pi t.$$

The proof of the theorem will be completed as soon as we show that D_N is a Dirac function (for N illimited) since

$$A_N(f)(x) = \int f(y)D_N(x-y)\,dy = \int f(x-t)D_N(t)\,dt \simeq f(x).$$

8.2.4 Lemma

For N illimited, the function $D_N(t)=(1/N)\sin^2 N\pi t/\sin^2 \pi t$ is a Dirac function.

8.2.5 Proof

Integrating term by term the finite sum

$$K_n(t)=e(t)^{-n}+\ \dots\ +1+\ \dots\ +e(t)^n$$

we find

$$\int K_n(t)\,\mathrm{d}t=1 \quad \text{for all } n\in\mathbb{N}.$$

Consequently, the average of the first N of these integrals is also normalized

$$\int D_N(t)\,\mathrm{d}t=1.$$

Moreover, let us see that we can choose a positive infinitesimal δ for which,

$$\int\limits_{|t|<\delta} D_N(t)\,\mathrm{d}t\simeq 1.$$

For this purpose, observe that

$$0\leq D_N(t)\leq \frac{1}{N\sin^2 \pi\delta} \quad \text{for } |t|\geq\delta.$$

Therefore,

$$\int\limits_{|t|\geq\delta} D_N(t)\,\mathrm{d}t\leq \frac{1}{N\sin^2 \pi\delta}\int\limits_{|t|\geq\delta}\mathrm{d}t\leq\frac{1}{N\sin^2\pi\delta}$$

and we shall have achieved our goal if we can make the last expression infinitesimal. It is enough to choose δ with $\delta^2=1/\sqrt{N}$. The choice $\delta=N^{-1/4}$ is suitable.

8.3 BERNSTEIN POLYNOMIAL APPROXIMATION

8.3.1 Bernstein interpolation

In this section, we shall study the approximation of continuous functions $f\colon[0,1]\to\mathbb{R}$ by polynomials. More precisely, we shall prove that the **Bernstein polynomials**

$$B_n(f)(x)=\sum_{0\leq k\leq n} f(k/n)\cdot\binom{n}{k}x^k(1-x)^{n-k}$$

converge uniformly to f.

Since the preceding statement is classical, it will be enough to establish it for standard f. In this case, by 3.4.7, it is enough to show that for illimited integers n, $(B_n f)(x) \simeq f(x)$ for all $x \in [0, 1]$. Thus, we shall fix the integer n and eventually consider the case n illimited.

8.3.2 A partition of unity

For a fixed integer n, consider the system of positive functions on $[0, 1]$

$$\phi_k(x) = \binom{n}{k} x^k (1-x)^{n-k} \quad (0 \leq k \leq n).$$

They constitute a *partition of unity* since the sum $\Sigma\,\phi_k$ is simply the binomial expansion of the nth power of $x+(1-x)=1$.

As a consequence, $B_n(f)(x)$ is a convex combination of the values $f(k/n)$ of f.

The reader should sketch the graph of these functions ϕ_k and in particular, verify that they have a maximum at $x=k/n$ (for $k=0$, $\phi_0(0)=1$ is maximal with resp to $x \in [0, 1]$). The derivative of ϕ_k at the origin also vanishes for $k \geq 2$ (whereas $\phi'_0(0)=-n$, and $\phi'_1(0)=n$). Of course,

$$\sum_{0 \leq k \leq n} \phi_k = 1 \Rightarrow \sum_{0 \leq k \leq n} \phi'_k = 0.$$

8.3.3 Monotonicity properties of the interpolation

It is clear on the definition that

$$f \geq 0 \Rightarrow B_n(f) \geq 0.$$

From this inequality and the linearity of B_n, we infer more generally

$$f \leq g \Rightarrow B_n(f) \leq B_n(g).$$

(apply B_n to the positive function $g-f$). We can still write this monotonicity of B_n in the useful form

$$|f| \leq g \Rightarrow |B_n(f)| \leq B_n(g)$$

(indeed, the assumed inequality is $-g \leq f \leq g$ to which B_n can be applied).

8.3.4 A few crucial cases

For the general approximation result to be proved presently, the special case of quadratic functions will be crucial. Thus, we start by studying it in some detail for $f_i(x)=x^i$, $i=0, 1, 2$.

We have already observed that when $f=f_0$ is the constant 1, $B_n f_0$ is the expansion of the nth power of $x+(1-x)$, hence is also the constant 1.

Compute now $B_n f_1$. Since

$$\frac{k}{n}\binom{n}{k}=\binom{n-1}{k-1},$$

we also see that $B_n f_1 = f_1$, so that $B_n f = f$ whenever f is a linear function. Finally, a few computations lead to

$$(B_n f_2 - f_2)(x) = \frac{1}{n}\, x(1-x).$$

Since $0 \le x(1-x) \le 1/4$ in the interval $I=[0, 1]$, we shall have

$$|(B_n f_2 - f_2)(x)| \le 1/(4n) \quad \text{for all } x \in I,$$

and this is infinitesimal for n illimited, but we shall need the more precise estimate just given.

8.3.5 Lemma

Let $\varphi_y(x)=(y-x)^2$. Then $B_n(\varphi_y)(y) \le 1/(4n)(y \in [0, 1])$.

8.3.6 Proof

Write $\varphi_y(x)=(y-x)^2 = y^2 - 2yx + x^2$

or

$$\varphi_y = y^2 \cdot f_0 - 2y \cdot f_1 + f_2$$

with the basic polynomial functions $f_i(x)=x^i$ $(0 \le i \le 2)$. Since φ_y vanishes at the point y

$$(B_n \varphi_y)(y) = (B_n \varphi_y)(y) - \varphi_y(y)$$

and we estimate the difference $B_n \varphi_y - \varphi_y$

$$B_n \varphi_y - \varphi_y = y^2 \cdot (B_n f_0 - f_0) - 2y \cdot (B_n f_1 - f_1) + (B_n f_2 - f_2).$$

Here, the explicit computations of 8.3.4 can be applied. They lead to the expected result since φ_y is a unitary second degree polynomial (the coefficient of x^2 in $\varphi_y(x)$ is 1).

8.3.7 Theorem (Bernstein)

Let f be an S-continuous bounded function on $I = [0, 1]$ with $\|f\|$ limited. Then $(B_n f)(x) \simeq f(x)$ for all $x \in I$ if n is illimited.

8.3.8. First proof

For a fixed point $y \in I$, we shall consider the function

$$f - f(y) = f - f(y) \cdot f_0.$$

which still satisfies all assumptions of the theorem. It will thus be enough to prove that

$$\|f\| \text{ limited, } f \text{ S-continuous at } y \text{ with } f(y)=0$$

imply

$$(B_n f)(y) \simeq 0 \quad \text{for } n \text{ illimited.}$$

For this purpose, if n is chosen (illimited), put $\delta = n^{-1/4}$, so that $\delta > 0$ is infinitesimal. Thus

$$|x-y| < \delta \Rightarrow x \simeq y \Rightarrow f(x) \simeq f(y) = 0$$

and

$$\sup_{|x-y|<\delta} |f(x)| = \varepsilon \text{ is still infinitesimal (cf. Lemma 3.4.5).}$$

We thus have a majoration of $|f|$ by

$$|f(x)| \leq \begin{cases} \varepsilon & \text{if } |x-y| < \delta \\ \|f\| & \text{if } |x-y| \geq \delta. \end{cases}$$

It is easy to replace this discontinuous majorant by a quadratic function g. Observing that

$$\varphi_y(x) = (x-y)^2 \geq \delta^2 \quad \text{when } |x-y| \geq \delta$$

we have

$$|f| \leq g = \varepsilon + \|f\| \cdot \varphi_y / \delta^2 \quad \text{everywhere on } I \text{ (cf. Fig. 8.1).}$$

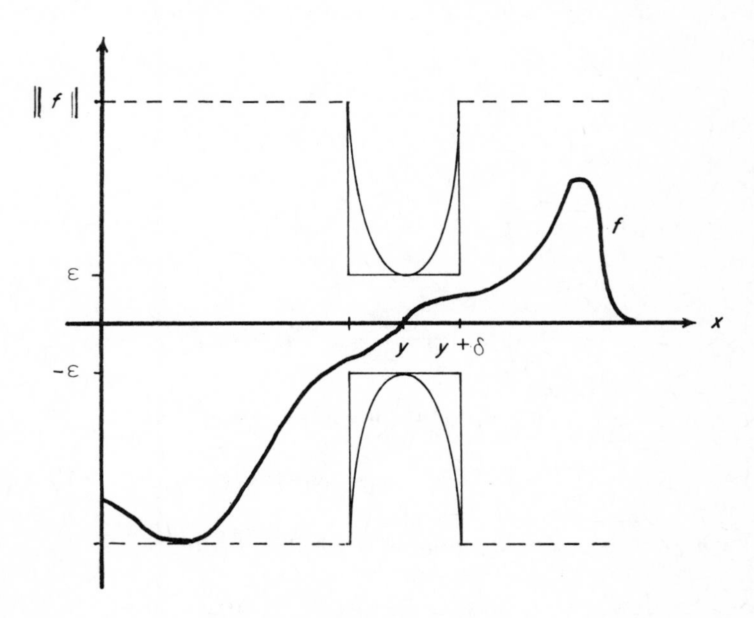

Fig. 8.1

Using the monotonicity property of the operator B_n, we shall also have

$$|B_n f| \le B_n g = \varepsilon + (\|f\|/\delta^2) \cdot B_n \varphi_y \simeq (\|f\|/\delta^2) \cdot B_n \varphi_y$$

and at the point y

$$|B_n f(y)| \le \varepsilon + (\|f\|/\delta^2) \cdot B_n \varphi_y(y) \le \varepsilon + \|f\|/(4\delta^2 n) \simeq 0$$

since our choice was precisely made to keep $\delta^2 n = \sqrt{n}$ illimited (recall that by assumption, $\|f\|$ is limited).

8.3.9 Second proof

Let us give another proof of Bernstein's theorem (closer to the original proof). Let us fix $x \in [0, 1]$ and write $p_k = \varphi_k(x)$. We estimate the sum

$$\sum \left(\frac{k}{n} - x\right)^2 p_k = (B_n f_2)(x) - 2x(B_n f_1)(x) + x^2$$

$$= x^2 + \frac{1}{n} x(1-x) - 2x \cdot x + x^2 = \frac{1}{n} x(1-x) \le 1/(4n) < 1/n.$$

When n is illimited, the main contribution to $\Sigma p_k = 1$ comes from the indices k for which k/n is close to x (the function φ_k indeed has a maximum at $x = k/n$). We shall make this assertion more precise now. Fix a positive σ and look at the set of indices

$$I_x = I_x(\sigma) = \{0 \le k \le n : |k/n - x| < \sigma\},$$

$$J_x = J_x(\sigma) = \{0 \le k \le n : |k/n - x| \ge \sigma\}.$$

Obviously

$$\sum_{J_x} (k/n - x)^2 p_k \ge \sigma^2 \sum_{J_x} p_k$$

and thus

$$\sum_{J_x} p_k \le \sigma^{-2} \sum_{J_x} (k/n - x)^2 p_k \le \sigma^{-2} \sum (k/n - x)^2 p_k < \frac{1}{\sigma^2 n}.$$

By choosing $\sigma = n^{-1/4} \simeq 0$ (since n is illimited) we shall have

$$\sum_{J_x} p_k \simeq 0 \quad \text{and consequently} \quad \sum_{I_x} p_k \simeq 1.$$

Now, for a given bounded, S-continuous f with $\|f\|$ limited,

$$\left| \sum_{J_x} f(k/n) p_k \right| \le \|f\| \sum_{J_x} p_k \simeq 0;$$

whereas

$$\sum_{I_x} f(k/n)p_k \simeq f(x)\sum_{I_x} p_k \simeq f(x),$$

since

$$k \in I_x \Rightarrow k/n \simeq x \Rightarrow f(k/n) \simeq f(x).$$

8.4. APPROXIMATION IN QUADRATIC MEAN

8.4.1 Parseval equality

Let f be a continuous 2π-periodic function $\mathbb{R}\to\mathbb{C}$ (or more generally any square summable function on the interval $[0, 2\pi]$). The quadratic norm of f and the Fourier coefficients of f are respectively defined by

$$\|f\|_2^2 = \frac{1}{2\pi}\int_0^{2\pi} |f(x)|^2\,\mathrm{d}x,$$

$$c_n = c_n(f) = (e_n|f) = \frac{1}{2\pi}\int_0^{2\pi} e^{-inx} f(x)\,\mathrm{d}x.$$

With these notations, Parseval's equality is

$$\|f\|_2^2 = \sum_{n\in\mathbb{Z}} |c_n|^2.$$

This equality will be proved in 8.4.4.

8.4.2 Best approximation lemma

With the preceding notations, for fixed N

$$\left\| f - \sum_{|n|\le N} a_n e_n \right\|_2 \quad \text{is minimal for } a_n = c_n = c_n(f).$$

This result is classical, well known, easy to establish . . . and we repeat its proof!

8.4.3 Proof

An elementary computation shows that

$$f - \sum_{|n|\le N} c_n e_n \quad \text{is orthogonal to } e_i \quad \text{for all } i \le N.$$

Pythagoras theorem can be applied to the orthogonal decomposition

$$f-\sum_{|n|\leq N} a_n e_n = \left(f-\sum_{|n|\leq N} c_n e_n\right)+\sum_{|n|\leq N}(c_n-a_n)e_n$$

furnishing

$$\left\|f-\sum_{|n|\leq N} a_n e_n\right\|_2^2 = \left\|f-\sum_{|n|\leq N} c_n e_n\right\|_2^2 + \sum_{|n|\leq N}|c_n-a_n|^2 \geq \left\|f-\sum_{|n|\leq N} c_n e_n\right\|_2^2$$

for any choice of coefficients $a_n(n\leq N)$. The equality holds precisely when $|c_n-a_n|=0$ for $n\leq N$, namely when $a_n=c_n$ for those values of n.

8.4.4 Proof of 8.4.1

We can assume that f is standard, 2π-periodic and continuous. (If f is only assumed square summable on $[0, 2\pi]$, there is a continuous f_1 with $\| f-f_1 \|_2 \simeq 0$, and also a continuous and periodic f_2 with $\| f_1-f_2 \|_2 \simeq 0$, so that f could be replaced by f_2 in the next proof.)

By 8.2.2 (Theorem of Fejer), there is a trigonometric polynomial

$$p=\sum_F a_n e_n$$

(where we can always assume that the finite sum extends over a set F of the form $|n|\leq N$, simply introducing more zero coefficients if necessary) with $p(x)\simeq f(x)$ for all $0\leq x\leq 2\pi$. Integrating the square of the modulus of the difference on this interval, we get

$$\left\|f-\sum_F a_n e_n\right\|_2^2 \quad \text{infinitesimal.}$$

A fortiori we shall have

$$\left\|f-\sum_F c_n e_n\right\|_2^2 \quad \text{infinitesimal}$$

by the best approximation Lemma 8.4.2 (with the same index set F). Evaluating the preceding square of norm (by scalar product of the vector with itself), we find after a couple of easy simplifications

$$0\leq \| f-\sum_F c_n e_n \|_2^2 = \| f \|_2^2 - \sum_F |c_n|^2 \quad \text{infinitesimal.}$$

Consequently, if f is standard, in which case the sequence $n \mapsto c_n$ is also standard

$$\sum_{n\in\mathbb{Z}} |c_n|^2 = \mathrm{st} \sum_F |c_n|^2 = \| f \|_2^2.$$

The convergence of $\Sigma_{n\in\mathbb{Z}} |c_n|^2$ is ensured since $\Sigma_F |c_n|^2 \leq \| f \|_2^2$ for all finite parts F.

8.5 EXERCISES

8.5.1 (a) Let D be a continuous Dirac function. Can you prove that $D(x) \simeq 0$ for all non infinitesimal x?

(b) Show that in the definition of a Dirac function, positivity is not essential, provided we assume that localization holds in the form

$$\text{there exists } 0 < \delta \simeq 0 \quad \text{with} \quad \int_{|x| \geq \delta} |D(x)| \, \mathrm{d}x \simeq 0.$$

8.5.2 Let E be a standard metric space. Then for $A \subset E$, consider the following conditions.

(i) A is dense in E,
(ii) for each standard $x \in E$, there is $a \in A$ with $d(a, x) \simeq 0$.

Show that they are not equivalent in general, but are equivalent if A is a standard part. Reformulate in a nonclassical way the density of the space of polynomials in $C([0, 1])$.

CHAPTER 9

DIFFERENTIAL EQUATIONS

9.1 REVIEW OF SOME CLASSICAL NOTIONS

9.1.1 Existence of solutions

Consider a differential equation of the first order given in the form

$$y' = f(x, y)$$

with a continuous function $f\colon \mathbb{R}^2 \to \mathbb{R}$. The *solutions* of such an equation are the differentiable functions $y = u(x)$ such that

$$u'(x) = f(x, u(x)).$$

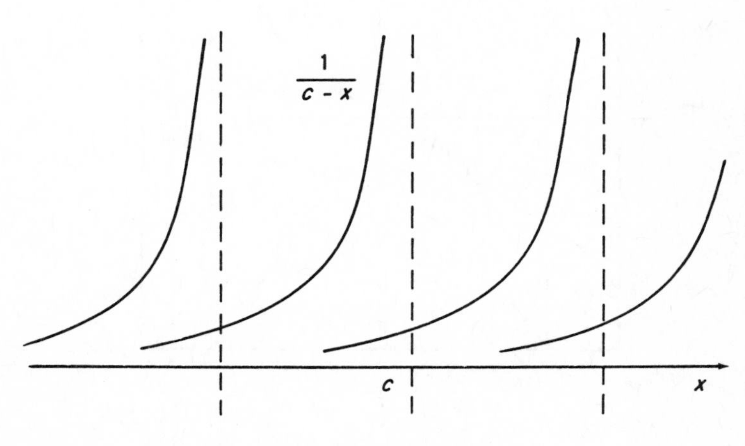

Fig. 9.1

As the example $y' = f(x, y) = y^2$ already shows, it is not sufficient to assume that the function f is defined and continuous everywhere to assure the existence of global solutions. In this example, the solutions are the functions $y = (c-x)^{-1}$ which have singularities (cf. Fig. 9.1). Any existence theorem must have a local character.

9.1.2 Construction of an appropriate interval

Let us fix a point $(a, b) \in \mathbb{R}^2$. We are looking for solutions of $y' = f(x, y)$ going through the preceding point, i.e. such that $u(a) = b$.

By continuity, the function f is bounded in the compact neighbourhood (a square)

$$V: |x-a| \leq 1, \qquad |y-b| \leq 1$$

of the initial point (a, b). Say $|f(x, y)| \leq M$ for all $(x, y) \in V$. It is clear that any curve going through the centre of the square V having a slope $\leq M$ (in absolute value) will have a part in the square for at least $|x-a| \leq 1/M$. More precisely (cf. Fig. 9.2), a portion of the curve will be contained in the rectangle

$$|x-a| \leq \inf(1, 1/M), \qquad |y-b| \leq 1.$$

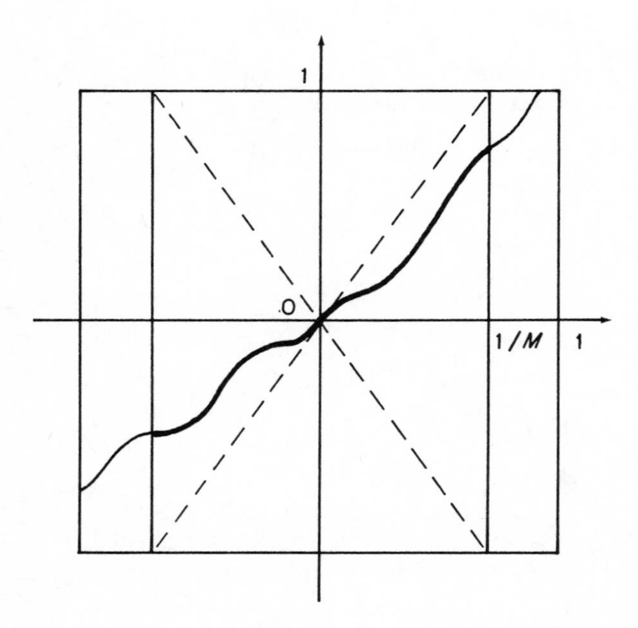

Fig. 9.2

9.1.3 Definition

A **safety rectangle** $R = I \times J \subset \mathbb{R}^2$ is a rectangle having diagonals with slopes majorizing $M = \max_R |f|$. Any curve $y = u(x)$ going through the centre of such a safety rectangle and having a slope $|u'(x)| \leq M$ must hit the vertical sides of R and thus be defined for all values of $x \in I$.

9.1.4 Uniqueness comment

It is quite remarkable that an existence theorem for solutions of $y' = f(x, y)$ can be established when f is continuous (cf. 9.2.1 below), even if this assumption is not sufficient to ensure uniqueness with respect to an initial data $u(a) = b$.

For example, the equation $y' = f(x, y) = 2|y|^{1/2}$ admits infinitely many solutions going through the origin: $u(0) = 0$. The solutions u_a defined by

$$u_a(x) = 0 \quad \text{for } x \leq a, \qquad u_a(x) = (x - a)^2 \quad \text{for } x \geq a$$

are indeed of the desired form when $a \geq 0$.

9.2 EXISTENCE THEOREM

9.2.1 Theorem (Peano)

Let $f: I \times J \to \mathbb{R}$ be a continuous function defined over a compact rectangle and satisfying

$$M = \max |f| \leq l(J)/l(I) \ (l \text{ denoting the length of an interval}).$$

Then there exists a solution $y = u(x)$ of the differential equation $y' = f(x, y)$ going through the centre of the rectangle, and defined for all values of $x \in I$.

9.2.2 Proof

Without loss of generality, we assume that the centre of the rectangle is at the origin $(0, 0) \in \mathbb{R}^2$, and we only construct a solution for positive values of $x \in I = [-a, a]$.

We start with the construction of an 'approximate' solution $y = v(x)$ as follows. Choose an infinitesimal $h > 0$. The formulas

$$v(0) = 0,$$
$$v((k+1)h) = v(kh) + f(kh, v(kh))h$$

define inductively the values of v at positive integral multiples (in I) of h. Extend v by a constant value on $[kh, (k+1)\, h[$:

$$v(x) = v([x/h]h) \quad \text{for positive } x \text{ in } I.$$

One can prove by induction

$$|v(kh)| \leq Mkh \quad \text{for } k \in \mathbb{N} \text{ with } kh \in I.$$

Hence for $0 \leq x \leq a$, we shall have

$$|v(x)| = |v(kh)| \leq Mkh \leq Mx \leq Ma \quad (\text{taking } k = [x/h]).$$

Since the statement of the theorem is classical, it is enough to prove it for standard data (transfer taking care of the generalization . . .). Thus, we assume that f (hence also I, M, a, . . .) is standard. The inequality just proved for $|v(x)|$ shows then that $v(x)$ is limited for $0 \leq x \leq a$. For x standard first, we define

$$u(x) = v(x)^*.$$

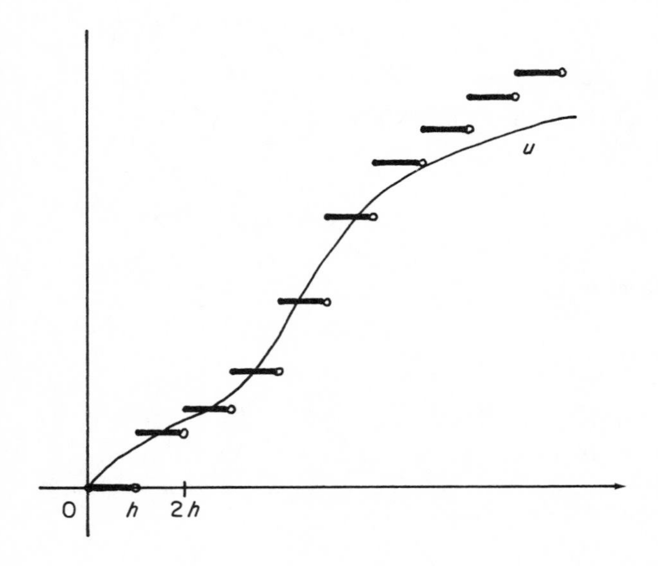

Fig. 9.3

This relation defines u implicitly for all x (there is only one standard function u taking the precedingly prescribed values at standard points x). The proof will successively show

- v is S-continuous.
- u is continuous,
- u is strictly differentiable and $u'(x) = f(x, u(x))$.

By induction, it is easy to see that

$$v(x) = \sum_{0 \leq kh < x} f(kh, v(kh)) \cdot h.$$

Thus, if $x < x'$:

$$v(x')-v(x)=\sum_{x \leq kh < x'} f(kh, v(kh))\cdot h$$

and

$$|v(x')-v(x)| \leq Mh \text{ (nb. of terms)} \leq Mh\left(1+\frac{x'-x}{h}\right)$$

$$= M(x'-x)+Mh.$$

Since $Mh \simeq 0$, S-continuity of the function v follows.

The continuous shade Theorem (4.3.5) shows that u is continuous.

Finally, take standard values for x and x' so that

$$u(x')-u(x)=\text{st}(v(x')-v(x))$$

and thus

$$|u(x')-u(x)| \leq \text{st}(M(x'-x)+Mh)=M(x'-x).$$

This classical inequality has standard parameters u and M, hence is still true for all $x < x'$ by transfer (thereby proving that u satisfies a Lipschitz condition). It is thus legitimate to take

$$x = (kh)^*, \qquad x'=kh \quad (k\in\mathbb{N} \text{ such that } kh\in I).$$

Thus $x \simeq x'$ and

$$\begin{aligned} u(kh) &\simeq u(x) \quad \text{(since } u \text{ and } x \text{ are standard)} \\ &= \text{st } v(x) \simeq v(x) \\ &\qquad\qquad \simeq v(kh) \quad \text{(since } v \text{ is } S \text{ continuous)} \end{aligned}$$

(observe that this sequence of inequalities replaces a classical proof using 3ε!). Uniform continuity of f in its second variable gives

$$A_k = f(kh, u(kh)) \simeq f(kh, v(kh)) = B_k$$

and averaging (cf. Proposition 3.1.4)

$$h \sum_{0 \leq kh < x} A_k \simeq h \sum_{0 \leq kh < x} B_k.$$

For standard x, we infer

$$u(x) \simeq v(x) = \sum_{\cdots} f(kh, v(kh))\cdot h \simeq \sum_{\cdots} f(kh, u(kh))\cdot h$$

and taking standard parts of the extreme terms

$$u(x) = \text{st} \sum_{0 \leq kh < x} f(kh, u(kh)\cdot h = \int_0^x f(t, u(t))\,\mathrm{d}t.$$

Since the parameters f and u of this equality are standard, it must remain true for all x (positive in I). Thus u is strictly differentiable (5.3) with $u'(x)=f(x, u(x))$ (cf. Proof 6.2.5).

The theorem is completely proved.

9.3 AN EXAMPLE

9.3.1 Construction of a particular differential equation

We intend to construct a standard differential equation $y'=f(x,y)$ for which the constructed solution (Proof 9.2.2) *depends* on the choice of infinitesimal $h>0$ chosen.

The standard, continuous function f is defined in $\mathbb{R}^2$ as follows:

$$f(x,y)=\begin{cases} 0 & \text{for } x\leq 0 \\ 4x & \text{if } x>0 \text{ and } y\geq \varphi_+(x) \\ -4x & \text{if } x>0 \text{ and } y\leq \varphi_-(x) \\ \text{if } x>0, & \text{linear in } y \text{ for } y\in[\varphi_-(x),\varphi_+(x)] \end{cases}$$

where

$$\varphi_\pm(x)=x^3\left(\pm\frac{1}{2}+\cos\frac{\pi}{x}\right) \quad \text{is defined for } x>0.$$

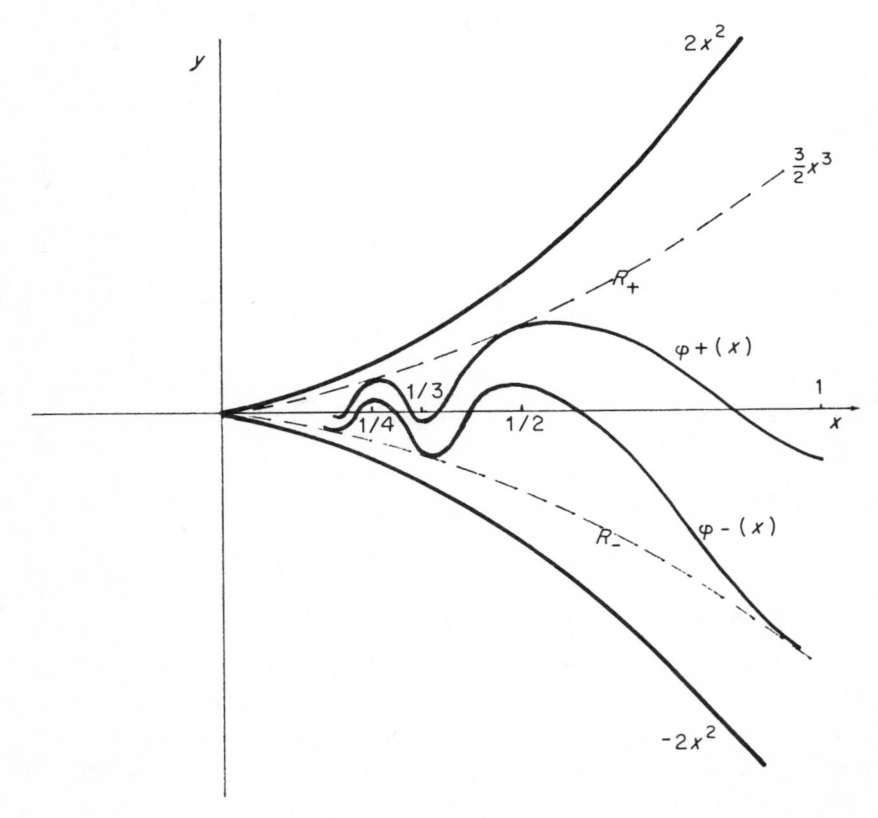

Fig. 9.4

Thus, for positive x, there are three regions of definition of f

R_+ defined by $y \geq \varphi_+(x)$ where the slopes are positive,
R_- defined by $y \leq \varphi_-(x)$ where the slopes are negative,
R transition region between the graphs of φ_+ and φ_-.

The transition region is situated between the graphs of the two functions $y = \pm\frac{3}{2}x^3$.

9.3.2 Families of solutions

Here are two continuous families of solutions of the differential equation $y' = f(x, y)$.

For $a \geq 0$, we have the family

$$y = \begin{cases} a & \text{for } x \leq 0 \\ a + 2x^2 & \text{for } x \in [10, 1] \end{cases}$$

and for a ≤ 0, the family

$$y = \begin{cases} a & \text{for } x \leq 0 \\ a - 2x^2 & \text{for } x \in [0, 1]. \end{cases}$$

We discover the two solutions satisfying $y(0) = 0$ (I do not know if these are the only ones . . .)

$$y = \begin{cases} 0 & \text{for } x \leq 0 \\ \pm 2x^2 & \text{for } x \in [0, 1]. \end{cases}$$

9.3.3 Dependence of choice of infinitesimal

Take now an infinitesimal $h > 0$ of the form $h = 1/k$ where k is an illimited odd integer. The approximate solution v goes through the points

$$P_0 = (0, 0), \qquad P_1 = (h, 0) \in R_+,$$

hence also through $P_2 = (2h, 4h^2)$, ... Let us show that all subsequent points remain in R_+. We even see that P_2 is situated above the graph of $y = (3/2)x^3$ hence also in R_+:

$$h < 1/3 \Rightarrow 4h^2 > 12h^3 = (3/2)(2h)^3.$$

Let us write

$$P_n = (nh, a_n h^2),$$
$$P_{n+1} = ((n+1)h, (a_n + 4n)h^2.$$

Then we can show that if P_n is in R_+, the same happens for P_{n+1}, at least if

$x(P_n) = nh < 2/3$:

$$a_n > (3/2)n^3h \Rightarrow a_{n+1} = a_n + 4n > (3/2)n^3h + 4n$$
$$a_{n+1} > (3/2)h(n^3 + 8n/3h).$$

Only remains to check

$$n^3 + 8n/3h > (n+1)^3 = n^3 + 3n^2 + 3n + 1$$

or

$$8n/3h > 3n^2 + 3n + 1, \qquad h < 8n/(9n^2 + 9n + 3)$$

which is obviously true! Consequently, the solution constructed in the proof 9.2.2 would agree with $2x^2$ on $[0, 2/3]$ hence also on $[0, 1]$ (there is only one solution of our system going through the point $x_0 = 2/3$ and $y_0 = 2(2/3)^2 = 8/9$).

Starting from an infinitesimal $h = 1/k$ where k is an illimited even integer, we would similarly see that $P_1 \in R_-$ and all subsequent points would also remain in R_-.

Thus, according to the nature of the infinitesimal $h > 0$, the solution with $u(0) = 0$ could be either $2x^2$ or $-2x^2$ (for $x \leq 1$).

CHAPTER 10

PERTURBATION OF A GREEN FUNCTION

10.1 GREEN FUNCTION

10.1.1 Review

We shall only consider the particular differential equation

$$y'' - \lambda^2 y = 0 \quad \text{(with a real parameter } \lambda > 0)$$

to which we refer by the initials ED. We are looking for solutions satisfying the conditions

$$y(0) = 0 \quad \text{and} \quad y \in L^2(0, \infty).$$

Let us recall that the Green function K of ED is the family of continuous functions K_ρ (or the function with two variables $K(r, \rho) = K_\rho(r)$) defined for $\rho > 0$ by the conditions

(a) K_ρ is a solution of ED except at most at $r = \rho$,
(b) $K_\rho(0) = 0$ and $K_\rho \in L^2$ $(0, \infty)$,
    (K_ρ satisfies the boundary conditions),
(c) K'_ρ is discontinuous at ρ, with a unit jump at this point. Moreover, the preceding conditions imply the following symmetry
(d) $K_\rho(r) = K(r, \rho) = K(\rho, r) = K_r(\rho)$.

The usefulness of this Green function is exhibited in the solution of the inhomogeneous equation

$$y'' - \lambda^2 y = f$$

satisfying $y(0)=0$ and $y \in L^2\ (0, \infty)$ in the form

$$y(\rho)=\int_0^\infty K_\rho(r) f(r) \quad \mathrm{dr}.$$

10.1.2 Computation of the Green function

As the solutions of ED are the linear combinations of exponentials $\exp(\pm\lambda r)$ (or of $Sh\ \lambda r$ and $Ch\ \lambda r$), we must have

$$K_\rho(r)=A\ Sh\ \lambda r \quad \text{for } r<\rho \quad \text{since } K_\rho(0)=0.$$

Similarly,

$$K_\rho(r)=B\mathrm{e}^{-\lambda r} \quad \text{for } r>\rho \quad \text{since } K_\rho \in L^2(0, \infty).$$

Continuity of K_ρ at $r=\rho$ gives the condition

$$A\ Sh\ \lambda\rho = B\,\mathrm{e}^{-\lambda\rho}$$

and the unit jump of K_ρ at $r=\rho$ leads to

$$-\lambda B\,\mathrm{e}^{-\lambda\rho}-\lambda A\ Ch\,\lambda\rho=1.$$

These two equations can be solved and lead to the values

$$A=-\lambda^{-1}\,\mathrm{e}^{\lambda\rho}, \qquad B=-\lambda^{-1}\ Sh\ \lambda\rho.$$

Thus, we have found

$$K_\rho(r)=\begin{cases} -\lambda^{-1}\,\mathrm{e}^{-\lambda\rho}\ Sh\ \lambda r & \text{for} \quad r<\rho \\ -\lambda^{-1}\ Sh\ \lambda\rho\ \mathrm{e}^{-\lambda r} & \text{for} \quad r>\rho. \end{cases}$$

10.2 NONSTANDARD PERTURBATION

10.2.1 Nonstandard differential equation

Let us now consider a perturbed differential equation ED′

$$y''+(V-\lambda^2)y=0$$

with the same limit conditions $y(0)=0$ and $y \in L^2\ (0, \infty)$. More precisely, we shall consider the case of a perturbation $V=V_\varepsilon$ of the following type

$$V_\varepsilon(r)=\begin{cases} a^2/\varepsilon^2 & \text{for} \quad r<\varepsilon \qquad (a>0) \\ 0 & \text{for} \quad r\geq\varepsilon \end{cases}$$

where $\varepsilon>0$ is infinitesimal: the support of the perturbation V is small, but the height of the perturbation itself is large. The Green function of this equation is still defined: in particular, the derivative K'_ρ is required to be continuous at $r=\varepsilon$.

10.2.2 Computation of the Green function

The form of the Green function for positive values of r is given in the following figure.

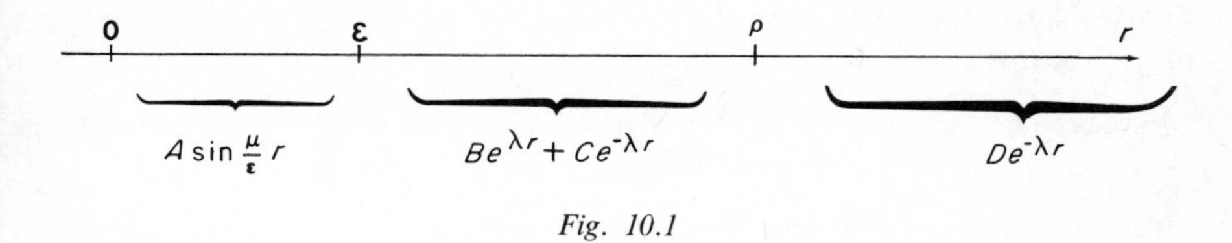

Fig. 10.1

We have introduced a constant μ satisfying $\mu^2/\varepsilon^2=a^2/\varepsilon^2-\lambda^2$ i.e.

$$\mu=\sqrt{(a^2-\varepsilon^2\lambda^2)}.$$

As before, the constants A, B, C, D depend on λ, ε and ρ. The continuity conditions can be written

(1) $A\sin\mu=B\,e^{\lambda\varepsilon}+C\,e^{-\lambda\varepsilon}$,
(2) $B\,e^{\lambda\rho}+C\,e^{-\lambda\rho}=D\,e^{-\lambda\rho}$,
(3) $A\mu\varepsilon^{-1}\cos\mu=B\lambda\,e^{\lambda\varepsilon}-C\lambda\,e^{-\lambda\varepsilon}$.

We still have to write the condition giving a unit jump of K'_ρ at $r=\rho$.

(4) $-D\lambda\,e^{-\lambda\rho}-B\lambda\,e^{\lambda\rho}+C\lambda\,e^{-\lambda\rho}=1$.

These four equations can be solved (the assiduous reader should check our computations . . .) and furnish

$$A=-e^{\lambda(\varepsilon-\rho)}/(\mu\varepsilon^{-1}\cos\mu+\lambda\sin\mu),$$
$$B=-e^{-\lambda\rho}/(2\lambda),$$
$$C=\frac{e^{\lambda}(2\varepsilon-\rho)}{2\lambda}\cdot\frac{\mu\cos\mu-\lambda\varepsilon\sin\mu}{\mu\cos\mu+\lambda\varepsilon\sin\mu},$$
$$D=C-e^{\lambda\rho}/(2\lambda).$$

10.2.3 Discussion

For this computation, we shall assume that the parameters a, λ and ρ are appreciable (i.e. limited and not infinitesimal), while $\varepsilon>0$ is infinitesimal. Since $\mu=\sqrt{(a^2-\varepsilon^2\lambda^2)}$, we see that $\mu\simeq a$.

For comparison, we recall that the Green function for $r<\rho$ in the non perturbed case is given by

$$-\frac{e^{-\lambda\rho}}{2\lambda}\cdot e^{\lambda r}+\frac{e^{-\lambda\rho}}{2\lambda}\cdot e^{-\lambda r},$$

and in the perturbed case by (for $\varepsilon < r < \rho$)

$$-\frac{e^{-\lambda\rho}}{2\lambda}\cdot e^{\lambda r}+\frac{e^{\lambda(2\varepsilon-\rho)}}{2\lambda}\cdot\frac{\mu\cos\mu-\lambda\varepsilon\sin\mu}{\mu\cos\mu+\lambda\varepsilon\sin\mu}\cdot e^{-\lambda r}.$$

The second term in this expression is obtained from the second term in the previous expression simply by multiplication by

$$e^{2\varepsilon\lambda}\cdot\frac{\mu\cos\mu-\lambda\varepsilon\sin\mu}{\mu\cos\mu+\lambda\varepsilon\sin\mu}\simeq\frac{\mu\cos\mu-\lambda\varepsilon\sin\mu}{\mu\cos\mu+\lambda\varepsilon\sin\mu}$$

This term describes the visible effect of the perturbation for r near ρ (or $r<\rho$). If $\mu\cos\mu$ is not infinitesimal, this term is still $\simeq 1$ and the perturbation has little effect. However, if

$$\mu\cos\mu\simeq 0,\quad \text{i.e. } \mu\simeq(n+1/2)\pi,$$

the fraction

$$\frac{\mu\cos\mu-\lambda\varepsilon\sin\mu}{\mu\cos\mu+\lambda\varepsilon\sin\mu}$$

can be far from 1! Indeed, this happens if we take the parameter a in the perturbation V_ε of the form $a=(n+1/2)\pi$, with $n\in\mathbb{Z}$ limited (i.e. standard). In this case

$$\mu^*=\text{st}(\mu)=a,$$

and more precisely

$$\begin{aligned}\mu&=a(1-\varepsilon^2\lambda^2/a^2)^{1/2}=a(1-\varepsilon^2\lambda^2/2a^2+\ldots)\\&=a-\varepsilon^2\lambda^2/2a+\ \ldots=(n+1/2)\pi-\varepsilon^2\lambda^2/2a+\ \ldots\end{aligned}$$

$$\mu\varepsilon^{-1}\cos\mu\simeq\pm\mu\varepsilon^{-1}\sin\varepsilon^2\lambda^2/2a\simeq\pm\mu\varepsilon\lambda^2\ 2a\quad\text{infinitesimal.}$$

The fraction

$$\frac{\mu\cos\mu-\lambda\varepsilon\sin\mu}{\mu\cos\mu+\lambda\varepsilon\sin\mu}\simeq\frac{-\lambda\sin\mu}{+\lambda\sin\mu}=-1$$

produces an observable perturbation at an appreciable distance from the origin.

The origin of this perturbation lies in the behaviour of the Green function in the first interval $[0, \varepsilon]$. In this part, we have

$$K_\rho(r)=A\sin\mu r/\varepsilon\quad\text{with } A=-\varepsilon\, e^{\lambda(\varepsilon-\rho)}/(\mu\cos\mu+\lambda\varepsilon\sin\mu)$$

so that

$$A\simeq 0\quad\text{when } \mu\cos\mu \text{ is not infinitesimal.}$$

In this case, the perturbation is small since its effect is to replace the condition $K_\rho(0)=0$ by $K_\rho(\varepsilon)\simeq 0$: this difference cannot be detected experimentally. If we assume that $a=(n+1/2)\pi$ so that $\mu\varepsilon^{-1}\cos\mu\simeq 0$, we shall have

$$A=-e^{\lambda(\varepsilon-\rho)}/(\mu\varepsilon^{-1}\cos\mu+\lambda\sin\mu)\simeq -e^{-\lambda\rho}/(\lambda\sin\mu)$$

appreciable and thus

$$K_\rho(\varepsilon) = A \sin \mu \simeq -e^{-\lambda\rho}/\lambda \quad \text{appreciable.}$$

The perturbation has now the visible effect of replacing the condition $K_\rho(0)=0$ (unperturbed case) by $K_\rho(\varepsilon) \simeq -e^{-\lambda\rho}/\lambda$ not infinitesimal.

10.3 ORIGIN OF THE PROBLEM

10.3.1 Schroedinger's equation

In physics, the wave function of a particle is a solution of the **Schroedinger equation**: this is a partial differential equation of the type

$$\mathscr{H}\Psi = -\frac{h}{i}\frac{\partial \Psi}{\partial t} = ih\frac{\partial \Psi}{\partial t}$$

where $\mathscr{H}$ denotes the Hamiltonian operator of the system (and h is Planck's constant divided by 2π). Let us assume that this Hamiltonian does not depend on time t and is simply given by

$$\mathscr{H} = -\frac{h^2}{2m}\Delta + V,$$

with a certain potential V, only depending on the space coordinates. The stationary solutions of Schroedinger's equation are those having the form $\Psi = e^{-i\omega t}\psi$ whence

$$ih\frac{\partial \Psi}{\partial t} = ih(-i\omega)\cdot\Psi = h\omega\cdot\Psi = E\Psi$$

since $h\omega = E$ is the energy of the particle. After a simplification by $e^{-i\omega t}$, we find the equation for the space part ψ of Ψ

$$\Delta\psi + 2mh^{-2}(E-V)\psi = 0.$$

With an attractive potential, bound states occur and we shall focus our attention to this case, namely

$$V = -h^2 V_\varepsilon/2m < 0, \quad E = -h^2\lambda^2/2m < 0.$$

The equation thus reduces to

$$\Delta\psi + (V_\varepsilon - \lambda^2)\psi = 0.$$

10.3.2 Spherical coordinates

It is well known that the Laplace operator in spherical coordinates is given by

$$\Delta\psi = \psi'' + 2r^{-1}\psi' + r^{-2}\cdot\Omega\psi \quad (\text{where } \psi' = \partial\psi/\partial r \ldots)$$

with a partial differential operator Ω only acting on the angular variables of ψ. Let us consider radial functions only, so that the Laplace operator is simply given by

$$\Delta\psi = \psi'' + 2r^{-1}\psi'.$$

In this case, the above Schroedinger equation reduces to

$$\psi'' + 2r^{-1}\psi' + (V_\varepsilon - \lambda^2)\psi = 0$$

or to

$$r\psi'' + 2\psi' + (V_\varepsilon - \lambda^2)r\psi = 0.$$

Since $r\psi'' + 2\psi' + (r\psi)''$, we shall put $r\psi = y$ and finally get the equation

$$y'' + (V_\varepsilon - \lambda^2)y = 0$$

precisely of the type considered in 10.2.1.

It is also interesting to translate the limit conditions for ψ in terms of y. Since the function ψ should be continuous at the origin, we have to impose

$$y(0) = r\psi(r)_{r=0} = 0$$

(more precisely, the continuity of ψ at the origin implies that $y(r) = \mathcal{O}(r)$ is of the order of r for $r \to 0$: this condition is automatically satisfied for the solutions of the differential equation satisfying $y(0) = 0$).

Finally, the condition $\psi \in L^2(\mathbb{R}^3)$ gives in spherical coordinates

$$\infty > \|\psi\|^2 = \iiint |\psi|^2 \cdot r^2 \,\mathrm{d}r \,\mathrm{d}\sigma = 4\pi \int_0^\infty |r\psi|^2 \,\mathrm{d}r = 4\pi \int_0 |y|^2 \,\mathrm{d}r.$$

This shows that $y \in L^2\,(0, \infty)$ and up to the factor 4π,

$$\psi \mapsto r\psi = y : L^2\,(\mathbb{R}^3) \to L^2(0, \infty)$$

is an isometry.

10.3.3 Nonstationary problem

It is interesting to observe that the resolution of the nonstationary problem also uses Green's function since

$$\partial\Psi/\partial t = -(i/h)\mathscr{H}\Psi$$

is solved by the evolution operator $U_t = \exp(-(i/h)\mathscr{H}t)$

$$\Psi(\mathbf{r}, t) = U_t \cdot \Psi(\mathbf{r}, 0).$$

This evolution operator can be computed by means of the functional calculus

$$f(\mathscr{H}) = \lim_{\varepsilon \to 0+} (2\pi i)^{-1} \int_{\mathbb{R}} (R_{\lambda + i\varepsilon} - R_{\lambda - i\varepsilon}) f(\lambda) \mathrm{d}\lambda,$$

where f denotes any analytic function on $\mathbb{R}$ (the exponential in our case) and R is the resolvant of the autoadjoint operator $\mathscr{H}$

$$R_\lambda = (\mathscr{H} - \lambda I)^{-1}.$$

This resolvant is precisely an integral operator having the Green function as kernel!

10.3.4 Final comment

Recall that the perturbation was produced by a potential of the form

$$V_\varepsilon(r) = \begin{cases} a^2/\varepsilon^2 & \text{for } r < \varepsilon \\ 0 & \text{for } r \geq \varepsilon \end{cases}$$

It is really interesting to observe that the perturbation produced by the short range potential V_ε is observable in spite of the fact that the measure $V_\varepsilon \, dx \, dy \, dz$ is infinitesimal. The potential V_ε has its support in the ball of radius ε (with volume $4\pi\varepsilon^2/3$) and its intensity is only of the order of $1/\varepsilon^2$. In particular V_ε is not a Dirac function: for any standard continuous function defined in a neighbourhood of 0 in $\mathbb{R}$, we have

$$\text{st} \iiint f \cdot dx \, dy \, dz = 0.$$

However, the perturbation is not negligible for certain limited values of the parameter a.

Ce qui a été compris n'existe plus!
(Paul Eluard)
[What has once been understood ceases to exist!]

CHAPTER 11

INVARIANT SUBSPACES

11.1 SITUATION OF THE RESULT

11.1.1 Historical comments

Continuous linear operators (we shall simply say 'operator') in Hilbert spaces have been extensively studied since the beginning of the century. In particular, attempts have been made to generalize results obtained in the finite dimension. However, it is still unknown if these operators always possess nontrivial closed invariant subspaces. For special classes of operators, such results can be established. In particular, Hermitian or compact operators have many such spaces. This is clear by Riesz theory for compact operators having a spectrum different from $\{0\}$ since any nonzero spectral value is an eigenvalue in this case (the corresponding eigenspace being finite dimensional). Compact operators having spectrum $\{0\}$ are also called **quasi-nilpotent** and also have nontrivial invariant subspaces: this has been shown in an article [1] by Aronszajn and Smith. The problem for operators having a square which is a compact operator was mentioned by Halmos in a series of open problems which remained open until the Bernstein–Robinson article [2] appeared. We shall give an internal translation of this paper here (i.e. using Nelson's approach of NSA).

Historically, the appearance of [2] was a first success of NSA: it showed that this theory is capable of solving unsettled conjectures. Specialists were of course puzzled by the method and Halmos himself translated the proof (with some help from Robinson) into classical terms. This lead him to publish [4] (both [2] and [4] appeared in the same issue of the *Pacific J. of Mathematics*!).

Although Robinson's proof is simpler and more conceptual than Halmos' (this is particularly true of the internal version that we are going to explain), it lost some interest when Lomonosov gave a simpler proof [9] using the

Leray–Schauder fixed point theorem. Hilden was even able to avoid recourse to this fixed point theorem and present a completely elementary classical proof of the same result (cf. [5]).

11.1.2 Bernstein–Robinson Theorem

Let H be an infinite dimensional Hilbert space and $T \in L(H)$ be an operator. If there exists a polynomial $p \neq 0$ such that the operator $p(T)$ is compact, then T has a nontrivial closed invariant subspace.

11.1.3 An example

Perhaps the simplest (trivial) example of a noncompact operator satisfying the assumptions of the Bernstein–Robinson Theorem is a projector. Indeed, projectors P are characterized by the property of being equal to their square: $P^2 = P$. The nonzero polynomial $p(X) = X^2 - X$ thus vanishes on all projectors, and thus $p(P) = 0$ is a compact operator! Obviously, projectors in spaces of dimension greater than 1 always have nontrivial invariant subspaces.

11.2 PRELIMINARY RESULTS

11.2.1 Review of a definition

Recall that if A is a subset of a standard metric space X, we can define a standard part of A by

$$A^* = {}^S\{x \in X : \text{there is an } y \in X \text{ with dist } (x, y) \simeq 0\}.$$

Instead of dist $(x, y) \simeq 0$, we shall sometimes write simply $x \simeq y$.

We shall have to use the basic facts

A^* is always closed,
if A is standard, then $A^* = \bar{A}$ contains A

(they were established in 3.3.5–3.3.6).

Moreover, it is clear that if H is a standard normed vector space, and $A = V$ is a vector subspace, then $A^* = V^*$ is also a vector subspace of H.

11.2.2 Lemma

For two subspaces $W \subset V$ of H

$$\dim V/W \leq 1 \Rightarrow \dim V^*/W^* \leq 1.$$

11.2.3 Proof

Assume that the codimension of W in V is ≤ 1 and take arbitrary standard elements x_1, $x_2 \in V^*$. Thus, there are $y_1 \in V$ with $x_1 \simeq y_i$. From the condimension hypothesis, there is a nontrivial linear relation $a_1y_1+a_2y_2 \in W$. Dividing the coefficients of the preceding relation by $a=a_i$ if $|a_i|=\max(a_1, a_2)>0$, we can assume that one $a_i=1$ and both are limited (more precisely, $|a_i|\leq 1$). Taking standard parts, we can write

$$a_1^*x_1+a_2^*x_2 \simeq a_1^*y_1+a_2^*y_2 \simeq a_1y_1+a_2y_2 \in W$$

This proves indeed

$$a_1^*x_1+a_2^*x_2 \in W^*.$$

Since this last relation is nontrivial (one of its coefficients is precisely 1), we see that the codimension of W^* in V^* is ≤ 1.

11.2.4 Proposition

Let H be a standard Hilbert space and $(e_i)\subset H$ any standard orthonormal basis. Then any near standard element y of H has infinitesimal components $y^i=(e_i|y)\simeq 0$ for illimited i.

11.2.5 Proof

Consider first a standard element $x\in H$. The sequence of components (x_n) of x is a standard sequence (in any standard orthonormal basis). Moreover, this sequence

$$i \mapsto x^i=(e_i|x)$$

tends to 0 for $i\to\infty$ by the Bessel inequality, or by the Parseval relation giving $\|x\|^2=\Sigma|x^i|^2$. Thus, $x^i\simeq 0$ for nonstandard i.

Recall now that by Definition 3.3.1, near standard elements $y\in H$ satisfy

$$\text{there exists a standard } x=y^*\in H \quad \text{with } y\simeq x.$$

The Cauchy–Schwarz inequality gives

$$|y_i-x_i|\leq \|y-x\|\simeq 0 \quad \text{for all } i.$$

Hence

$$y_i\simeq x_i \quad \text{for all } i \quad \text{and} \quad y_i\simeq x_i\simeq 0 \quad \text{for nonstandard } i.$$

11.2.6 Application

The preceding proposition will be used as follows. Let us take a standard compact operator T in H. Then, the image of the unit ball $B_1=B_1(0)\subset H$ must be contained in a standard compact subset of H. It follows that every element Tx (where $\|x\|\leq 1$ or more generally $\|x\|$ limited) is near standard. Thus, we have

$$\begin{aligned} x \text{ limited} &\Rightarrow Tx \text{ near standard} \\ &\Rightarrow (Tx)^i=(e_i|Tx)\simeq 0 \quad \text{for nonstandard } i. \end{aligned}$$

Taking in particular the unit basis vector $x = e_j$, we see that the matrix coefficients $a^i_j = (e_i|Te_j)$ of T in the basis (e_i) satisfy

$$a^i_j \simeq 0 \quad \text{for } i \text{ nonstandard } \textit{and all } j.$$

In other words, lines with nonstandard index i (in the matrix of T) only have infinitesimal elements (observe that this property really depends on the fact that the operator T is compact: the identity operator has identity matrix with one 1 in each line, hence fails to have this property!).

11.3 PROOF OF THE BERNSTEIN–ROBINSON THEOREM

11.3.1 Preliminary reductions

Since the statement of the theorem is classical, it is enough to prove it for standard data (transfer taking care of the generalization). Thus, we shall assume that the operator T and the polynomial $p \neq 0$ (for which $p(T)$ is compact) are standard. This also implies that the Hilbert space H in which T is defined is standard and if we write

$$p_n(X) = a_n X^n + \ldots + a_0$$

the degree n and all coefficients a_i of p are standard. Let us choose a nonzero standard $e \in H$. The smallest closed subspace of H containing e and all its transforms $T^i e$ is obviously standard, invariant under T, and separable. Thus we are reduced to the case of a separable Hilbert space H.

Since we are assuming $p \neq 0$, its degree n is well defined and we can assume that $a_n \neq 0$. Dividing throughout by this coefficient we can assume that p is unitary, still standard, hence with limited coefficients.

If $p(T) = 0$, the finite dimensional space generated by the $T^i e$ for $0 \leq i \leq n = \deg p$ is obviously invariant under T and we are reduced to the finite dimensional case. Thus we could assume that $p(T) \neq 0$.

In any case, let us choose and fix a standard normed $e \in H$ with $\{T^i e : i \in \mathbb{N}\}$ total in H. The sequence

$$i \mapsto T^i e$$

is thus standard.

11.3.2 First part of the proof of 11.1.2

We shall construct a finite dimensional (hence closed) subspace V in H which is nearly invariant under T in the following sense. Denote by P the orthogonal projector from H to V and let P_V be the restriction of PT to V (we consider that P_V is an operator in V). The construction will be such that

$$T_V x \simeq Tx \quad \text{for all limited } x \in V.$$

Let us first use the Gram–Schmidt orthonormalization process to deduce from the sequence

$$e,\ Te,\ T^2e,\ \ldots$$

a standard orthonormal basis (e_k) of H with

$$e_0 = e, \qquad e_k = \text{linear combination of the } T^ie \text{ for } i \le k$$

(recall that we assume that H is infinite dimensional, topologically generated by the linearly independent family (T^ie)). The coefficient of T^ke in e_k is nonzero.

By construction of the basis (e_i), the matrix A of T in this basis will nearly be in upper triangular form

$$a^i_j = 0 \quad \text{for } i > j+1 \quad (\text{instead of } i > j!).$$

By induction, it is easy to show that the powers T^n will have matrices

$$A^n = (a^i_j(n)) \quad \text{with } a^i_j(n) = 0 \quad \text{for } i > j+n.$$

Line by column matrix multiplication also shows that the coefficients in this first nonzero diagonal are simply given by

$$a^{j+n}_j(n) = \prod_{j \le i \le j+(n-1)} a^{i+1}_i.$$

The matrix $B = (b^i_j)$ of the operator $p(T)$ will inherit the preceding property in the following form

$$b^i_j = 0 \quad \text{for } i > j+n \quad \text{and} \quad b^{j+n}_j = a_n \prod_{\ldots} a^{i+1}_i.$$

But the operator $p(T)$ is compact, hence must have infinitesimal coefficients in nonstandard rows by 11.2.6. In particular,

$$b^{j+n}_j \simeq 0 \quad \text{for } j \text{ illimited}.$$

Since a_n is standard and nonzero, it implies that

$$\prod_{j \le i \le j+(n-1)} a^{i+1}_i \simeq 0 \quad \text{for illimited } j.$$

In this product of n (standard) terms, at least one must be infinitesimal. We can thus find an illimited $\nu(\ge j+1)$ with $a^\nu_{\nu-1} \simeq 0$ and the finite dimensional space V generated by the e_i for $i < \nu$ has the expected property, as we now show. For x limited and in V, the last component $x^{\nu-1}$ of this vector must also be limited (by the Cauchy–Schwarz inequality) and

$$\begin{aligned} Tx &= T\left(\sum_{i<\nu-1} x^ie_i + x^{\nu-1}e_{\nu-1}\right) \\ &= PTx + x^{\nu-1}a^\nu_{\nu-1}e_\nu \simeq PTx = T_Vx. \end{aligned}$$

By nonstandard induction, we see that

$$T^k x \simeq (T_V)^k x \quad \text{for all standard integers } k,$$

and in particular, we infer that

$$p(T)x \simeq p(T_V)x \quad \text{for limited } x \in V.$$

The preceding construction can be illustrated by a drawing of the matrices A of T and resp. A_V of T. The last is the square block defined by i and $j<v$ in the first. It contains all nonzero elements of the first v columns of A *except* the coefficient a^v_{v-1} just situated below the far end corner of A_V.

$$\left|\begin{array}{cccccc}
a^0_0 & a^0_1 & \ldots\ldots & a^0_{v-1} & a^0_v & \ldots \\
a^1_0 & a^1_1 & \ldots\ldots & a^1_{v-1} & a^1_v & \ldots \\
0 & a^2_1 & & & & \\
0 & 0 & \ldots a^{v-1}_{v-2} & a^{v-1}_{v-1} & a^{v-1}_v & \ldots \\
\hline
0 & 0 & \ldots\ 0 & a^v_{v-1} & a^v_v & \ldots \\
0 & 0 & \ldots\ 0 & 0 & & \ldots \\
 & & \ldots & & & \ldots \\
0 & 0 & \ldots\ 0 & 0 & & \ldots
\end{array}\right|$$

Fig. 11.1

11.3.3 Second part of the proof

By the Jordan theorem for the finite dimensional operator T_V we can find a filtration of V by an increasing sequence of subspaces

$$\{0\} = V_0 \subset V_1 \subset \ldots \subset V_v = V$$

having the properties

$$\dim(V_i) = i,$$

$$V_i \quad \text{is invariant under } T_V.$$

Still call P_i the orthogonal projector from H onto V_i (hence $P_v = P$).

The standard subspaces

$$H_i = V_i^* = {}^S\{x \in H : \exists y \in V_i \quad \text{with } x \simeq y\}$$

of H will now play a crucial role. Obviously

$$\{0\} = H_0 \subset H_1 \subset \ldots \subset H_v = H.$$

I claim that these closed subspaces H_i of H are *invariant* under T. Take indeed a standard $x \in H_i$, say $x \simeq y \in V_i$. By the best approximation property of orthogonal projections, we shall *a fortiori* have $x \simeq P_i x$ and S-continuity of T at the origin gives

$$\begin{aligned} Tx &\simeq TP_i x \\ &\simeq T_V P_i x \quad \text{since } P_i x \text{ is limited,} \end{aligned}$$

and $T_V P_i x \in V_i$ since V_i is invariant under T_V. This has shown that

for each standard $x \in H_i$, Tx also belongs to H_i.

My claim is proved by transfer since H_i is standard.

11.3.4 Third part of the proof

Let us now introduce the positive real numbers

$$d_i = \|p(T_V)P_i e\|.$$

From $P_0 e \in V_0 = \{0\}$, we see that $d_0 = 0$ and at the other extreme,

$$P_v e = e \in V \Rightarrow$$

$p(T)e$ is standard and a linear combination of $e_0, \ldots e_n = p(T_V)e \in V$ hence

$$d_v = \|p(T_V)\mathrm{e}\| \quad \text{is standard and} > 0.$$

Question: When can H_i be $\{0\}$?
Answer: Only when $d_i \simeq 0$.

Indeed, $P_i e$ is limited in V hence $p(T_V)P_i e \simeq p(T)P_i e$ is near standard (because $p(T)$ is compact!), say

$$p(T_V)P_i e \simeq x \quad \text{standard in } H_i = \{0\}.$$

Thus

$$p(T_V)P_i e \simeq 0 \quad \text{and} \quad d_i \simeq 0.$$

Question: When can H_i be equal to H?
Answer: Only when $d_i \simeq d$.

Indeed, if $e \in H_1$, we can find $y \in V_i$ with $e \simeq y$ and again by the best approximation property of orthogonal projections $e \simeq P_i e$. This implies $p(T_V)e \simeq p(T_V)P_i e$, whence $d = d_v \simeq d_i$.

The proof of the Bernstein–Robinson theorem can now be concluded by exhibiting a nontrivial closed invariant subspace among the H_i. Certainly H_i would be nontrivial if we could prove that $0 < d_i^* < d$.

Consider the smallest integer j with $d_j > d/2$:

$$d_{j-1} \leq d/2 \Rightarrow d_{j-1} \quad \text{not} \simeq d \Rightarrow H_{j-1} \neq H.$$

Thus,

- either $H_{j-1} \neq \{0\}$ (and the proof is finished!)
- or $H_{j-1} = \{0\}$ (and $d_{j-1} \simeq 0$).

In the second case,

$$d_j > d/2 \Rightarrow d_j \quad \text{not} \simeq 0 \Rightarrow H_j \neq \{0\}.$$

But by Lemma 11.2.2, dim $H_j = 1$ in this case, and we have again found a nontrivial closed invariant subspace of T among the H_i.

11.4 COMMENTS

11.4.1 Back to the proof of the Bernstein–Robinson Theorem

In the second part 11.3.3 of the preceding proof, the spaces H_i are standard for all indices i. But we have not defined a map $i \mapsto H_i$ since the non classical relation (i, H_i) is not classical (and not set forming). It cannot be used to define a graph and the situation should be compared with that of defining the standard part x^* of a real number $x \in [0, 1]$. The space H_j is well defined for each i, but there is no map $i \mapsto H_i$! However, $i \mapsto d_i$ *is* a map since it is classically defined, with certain (nonstandard) parameters T_V and V_i. It would have been possible to work with the distances $d_i = \|p(T)P_i e\|$, but even here P_i is not necessarily standard when i is!

11.4.2 An example

The construction of 11.3.3 can be illustrated in the example of operator T defined in

$$l^2\,(\mathbb{N}) = \text{space of square summable sequences } \mathbb{N} \rightarrow \mathbb{C}$$

by

$$T(e_i) = e_{i+1}/(i+1) \quad \text{for } i \geq 0$$

(e_i denotes the sequence having only one nonzero element equal to 1 at place i: e_i is the characteristic function of the point i).

This operator is compact with spectrum $\{0\}$: it is quasi-nilpotent. Since T is already compact, we can take the polynomial $p(X) = X$ here.

The truncation

$$V = l^2([0, \nu[) \quad (\text{with dim } V = \nu)$$

is adequate for T as soon as ν is illimited, and

$$T_V e_i = Te_i \quad \text{if } i < \nu - 1,$$
$$T_V e_{\nu-1} = 0 \quad (Te_{\nu-1} = e_\nu/\nu \simeq 0).$$

The Jordan filtration of V is given by the

$$V_i = l^2([\nu - i, \nu[).$$

Consequently, in this case,

$$\nu - i \text{ illimited} \Leftrightarrow H_i = \{0\},$$

$$\nu - i \text{ limited} \Leftrightarrow H_i = l^2([\nu - i, \infty[).$$

Since $e = e_0$ (as in the proof), $P_i e \in V_i = l^2([\nu - i, \nu[)$.

$$P_i e = 0 \quad \text{for } \nu - i > 0 \quad (\text{i.e. } i < \nu).$$

Hence, $d_i = 0$ for $i = \nu$ whereas $d_\nu = d = 1$. The idea of 11.3.4 is to look at $H_{\nu-1} = l^2([1, \infty[)$. This is indeed a closed, proper invariant subspace for T!.

INDICATIONS FOR THE EXERCISES

CHAPTER 1

1.9.1 Forget that n is nonstandard and try to prove the result for all integers n.

1.9.2 It is easy to construct set forming relations corresponding to the empty set.

1.9.3 If you have the impression that one of the (a), (b) or (c) parts is true, you are making a mistake . . . .

For (a), consider the set of even numbers,
for (b), consider the set of prime numbers,
for (c), consider the numbers 2^m.

CHAPTER 2

2.8.1 Compare the standard sets E and $\varnothing$.

2.8.2 Take a finite part F containing all standard elements of E . . . (and use 2.8.1 to answer the second question).

2.8.3 If E (or F) is empty, the Cartesian product is empty. Prove

$$E \times F \text{ standard } \textit{and} \text{ not empty} \Rightarrow E \text{ and } F \text{ standard.}$$

2.8.4 Consider the standard set

$$A = {}^{S}\{n \in \mathbb{N}: \mathrm{P}(n)\}.$$

2.8.5 The sets B and C are standard, B and A have the same standard elements . . . and $\varepsilon/2 \in \mathrm{A}$ but $\varepsilon/2 \notin C$.

2.8.6 Compare the standard subsets of $\mathbb{N} \times \mathbb{N}$

$$^S\{(x, y): R(x, y)\} \quad \text{and} \quad \Delta = \{(x,x): x \in \mathbb{N}\} \text{ (diagonal).}$$

2.8.7 For any function $f: E \to E$, we can form the subset

$$\{x \in E: f(x) = x\} \subset E$$

(this is just set theory, nothing to do with NSA).

2.8.8 (a) If E has only one element, this element is either standard or nonstandard. Proceed by (nonstandard) induction using (2.8.4).

(b) Consider the set $^S\{x \in E: P(x)\}$.

(c) If x is nonstandard, E and $E - \{x\}$ have the same standard elements.

2.8.9 Use transfer.

CHAPTER 3

3.5.1 Take $n \in \mathbb{N}$ nonstandard and consider $\varepsilon = 1/n!$.

3.5.2 Even if $\varepsilon > 0$ is rational, it is not always possible to find $n \in \mathbb{N}$ with $n\varepsilon$ standard and > 0.

3.5.3 (a) Consider a suitable linear function $f(x) = kx$.

(b) Consider the set of integers $n \in \mathbb{N}$ for which

$$|f(x)| \leq 1/n \quad \text{for all } |x| \leq n$$

and use principle 2.4.4.

3.5.4 If $A \in \mathbb{N}$ is nonstandard, we obviously have $|f(x)| < A$ (for all x).

3.5.5 Use transfer to prove (i) $\Rightarrow$ (ii) and the preceding exercise for (iii) $\Rightarrow$ (i).

3.5.6 Is there a set containing precisely the real numbers x with $x^* = 0$?

3.5.7 Forget first that ε is infinitesimal. Consider the set $f^{-1}(0)$ to answer the second question.

3.5.8 Let x be the number in question. Can you compute the square of x^*?

3.5.9 Consider a set of the form $[\varepsilon, 1]$.

3.5.10 When is the sequence $n \mapsto (1 + x/n)^n$ standard?

3.5.11 Consider constant sequences $a_n = \varepsilon$ (with $a = 0$). Also consider sequences of the form $a_n = A/n$ where A is illimited.

3.5.12 Consider only the case of standard sequences and use the characterization of convergence given in Theorem 3.4.1.

3.5.14 (b) For each standard $\varepsilon > 0$, we can write (in the real case)

$$(1-\varepsilon)\, a_i \leq b_1 \leq (1+\varepsilon)\, a_i$$

and one can sum these inequalities.

3.5.20 The set U_A is the complement of $(X-A)^*$ (both sets are standard so that the equality $X - U_A = (X-A)^*$ is checked on standard elements).

SOLUTIONS OF THE EXERCISES

CHAPTER 1

1.9.1 The indicated result is classical when formulated for all initial intervals $I_n \subset \mathbb{N}$. It is proved by (classical) induction on n. When proved for all $n \in \mathbb{N}$, can anyone doubt its validity for nonstandard n? (When a result $P(n)$ is proved for all integers $n \in \mathbb{N}$, should we have any doubts saying that $P(p)$ is true for a particular prime p?)

1.9.2 The property $P(x)=$'x is standard and x is nonstandard' is nonclassical and *is* set forming (in any set E): it defines the empty set $\varnothing$.

1.9.3 (a) The set of even integers is infinite (this is known!) hence contains a nonstandard element: let $n=2m$ be a nonstandard even number. This integer admits the standard prime divisor 2.

(b) The set of primes is infinite (known since Euclid!) hence contains a nonstandard element. Let p be a nonstandard prime. This integer has only one prime factor and $k=1$ is standard in this case.

(c) Take a nonstandard $m \in \mathbb{N}$ and consider $n = 2^m > m$. Thus n is also nonstandard. The only prime factor of n is the standard prime 2 ($k=1$ in this case, whence another example for (b)!). One could also start with the infinite set consisting of the powers of 2 and take a nonstandard element in this set . . . .

CHAPTER 2

2.8.1 Recall that two standard sets are equal as soon as they have the same standard elements (transferred extensionality 2.2.4). In particular: a standard set having no standard element is empty (the empty set is standard).

2.8.2 The complement C of a finite subset F of the infinite set E is infinite. When F contains all standard elements, C can only contain nonstandard elements. By 2.8.1 above, C cannot be standard.

2.8.3 The two projections $p_E: E \times F \to E$ and $p_F: E \times F \to F$ are surjective *as soon as* $E \times F$ is not empty, i.e. as soon as both E and F are not empty. If $E \times F$ is standard and not empty, take a standard element $(e,f) \in E \times F$. Then e and f are standard and e.g. $E = \{x : (x,f) \in E \times F\}$ must be standard since it is uniquely characterized by a classical formula containing two parameters f and $G = E \times F$ having standard values.

2.8.4 The standard set $A = {}^S\{n \in \mathbb{N}: P(n)\}$ coincides with $\mathbb{N}$ as we see by the following argument. The classical formula

$$\forall^s n (n \in A \Rightarrow n+1 \in A)$$

gives by transfer (the parameter A has a standard value)

$$\forall n\, (n \in A \Rightarrow n+1 \in A).$$

Since we are also assuming that $0 \in A$, a classical induction shows that $A = \mathbb{N}$. Finally, only the standard elements of A are characterised by the property P: we can only conclude that $P(n)$ is true for standard n.

2.8.5 We have $B = C \subset A$ with $c \neq A$ (thanks to $\varepsilon/2$). The set A is not standard since its upper bound is not standard (A and B have the same standard elements and are distinct, hence cannot be both standard).

2.8.6 The two sets given in the indication to this exercise are standard and have the same standard elements, hence coincide.

2.8.7 We know that there is no subset of $[0, 1]$ consisting precisely of the standard elements. Consequently, there is no function f with $f(x) = x^*$ (the condition $x = x^*$ precisely characterizes the standard elements in the given interval). Let us repeat several arguments for the nonexistence of a set containing precisely the standard elements of $I = [0,1]$. If a subset A consists only of standard elements, it must be standard and finite; but if A also contains all standard elements, it must coincide with I (since it is standard). Variant: if there was a subset A of I consisting precisely of the standard elements, we could define a subset

$$\{n \in \mathbb{N} : 1/n \in A\} \subset \mathbb{N}$$

consisting precisely of the standard integers, but this is impossible since there is no smallest nonstandard integer.

2.8.8 (a) We can use nonstandard induction as mentioned in the indication for this exercise (one could even start with the empty set, in which *any* relation is set forming!). To prove the induction step for a standard cardinal n (hence $n+1$ also standard) one just uses the fact that an element has to be standard or nonstandard ('excluded middle third principle').

(b) In a standard finite set, all elements are standard and in ${}^S\{x \in E : P(x)\}$, *all* elements are distinguished by the property P.

(c) The sets E and $E - \{x\}$ are distinct although they have the same standard elements. Hence one of them at least must be nonstandard. Since E is standard, $E - \{x\}$ must be nonstandard.

2.8.9 If a and $b \neq 0$ are standard, so is a/b (this number is uniquely defined by a standard formula having parameters a and b, assuming standard values in our case).

Conversely, if $x \in \mathbb{Q}$ there is a unique reduced representation as given in the exercise: b is uniquely defined classically from x and will be standard if x is (then $a = bx$ will also be standard).

CHAPTER 3

3.5.1 If n is an illimited integer, $n!$ is divisible by all standard integers. Take any standard rational $a/b \neq 0$ written in reduced form (as in Exercise 2.8.9) with b standard. Thus a/b is an integral multiple of $1/n!$.

3.5.2 Take a nonstandard prime integer p (the set of primes is infinite) and consider $\varepsilon = p^{-1} + p^{-2} > 0$. This number is infinitesimal (e.g. since $\varepsilon < 2p^{-1} = 2/p$). Let us find the integers $n \in \mathbb{N}$ for which $n\varepsilon$ is standard. In this case

$$\begin{aligned} n/p < n/p + n/p^2 = n\varepsilon \quad &\text{is standard} \\ \Rightarrow n/p \quad &\text{limited} \\ \Rightarrow n/p^2 = p^{-1}(n/p) &\simeq 0. \end{aligned}$$

But

$$n(p+1)/p^2 = n\varepsilon \quad \text{standard}$$

and this quotient representation cannot be *the* reduced one since here the denominator p^2 is nonstandard:

$n(p+1)$ and p^2 are not relatively prime
$\Rightarrow p$ must divide $n(p+1)$
$\Rightarrow p$ must divide $n(p$ and $p+1$ are relatively prime)
$\Rightarrow n = n'p$ for some $n' \in \mathbb{N}$
$\Rightarrow n'(p+1)/p = n\varepsilon$ standard
$\Rightarrow n'(p+1)$ and p are not relatively prime (as before)
$\Rightarrow p$ must divide n' (as before), say $n' = n''p$
$\Rightarrow n''(p+1) = n\varepsilon$ standard
$\Rightarrow n'' = 0$ (and thus $n' = 0$, $n = 0$).

This shows that the only standard multiple of ε is 0.

3.5.3 (a) Take an infinitesimal $\varepsilon > 0$ and consider the function $f(x) = \varepsilon x$.

(b) The set $A \subset \mathbb{N}$ defined in the indication for this exercise contains all standard integers n. Since there is no set containing precisely the standard integers, this set contains a nonstandard integer ν and we must hence have $f(\nu)$ infinitesimal since

$$|f(\nu)| \leq 1/\nu \simeq 0.$$

3.5.4 and **3.5.5** Obvious with the indications.

3.5.6 There is no set containing precisely the infinitesimal real numbers (otherwise there would be a largest infinitesimal of the form $1/n$, corresponding to a minimal nonstandard $n \in \mathbb{N}$). Consequently, there is no function on $[-1, 1]$ satisfying $f(x) = x^*$ (no standard or nonstandard f would do).

3.5.7 The functions f_ε are well defined for all $\varepsilon > 0$. In particular, they are well defined when ε is infinitesimal. If f_ε is a standard function, the set $f_\varepsilon^{-1}(0) = [0, \varepsilon[$ is standard and thus

$$\varepsilon = \sup\, [0, \varepsilon[\quad \text{is standard.}$$

If $\varepsilon > 0$ is infinitesimal, f_ε cannot be standard.

3.5.8 Using the rule $(xy)^* = x^*y^*$ for $x = y$, we infer

$$(x^*)^2 = (x^2)^* = (2 + \varepsilon)^* = 2.$$

Hence $x^* = \sqrt{2}$ (note that $x \geq 0 \Rightarrow x^* \geq 0$). Another way of proceeding would consist in guessing the answer and proving it by estimating the difference $\sqrt{(2+\varepsilon)} - \sqrt{2}$ in a more classical way

$$\sqrt{(2+\varepsilon)} - \sqrt{2} = \frac{(\sqrt{(2+\varepsilon)} - \sqrt{2})(\sqrt{(2+\varepsilon)} + \sqrt{2})}{\sqrt{(2+\varepsilon)} + \sqrt{2}} = \frac{(2+\varepsilon) - 2}{\sqrt{(2+\varepsilon)} + \sqrt{2}}$$

whence for example

$$\sqrt{(2+\varepsilon)} - \sqrt{2} \leq \varepsilon/\sqrt{2} \leq \varepsilon \simeq 0.$$

3.5.9 If $\varepsilon > 0$ is infinitesimal, $\varepsilon^* = 0 \notin [\varepsilon, 1]$ although this set is closed. Even more simply, take the set $A = \{\varepsilon\}$.

3.5.10 The sequence $n \mapsto (1 + x/n)^n$ is standard when x is standard. Use Theorem 3.4.1 since the limit is known to be e^x.

3.5.11 The condition (i) suggests a movement of the a_n towards a whereas condition (ii) is purely static! This interpretation shows how to fulfill each of them separately (if we are not required to exhibit standard sequences).

3.5.12 By transfer, it is enough to prove the result for standard sequences, and then it is possible to use condition (ii) of Theorem 3.4.1. To simplify the problem, subtract the constant standard value a from each a_n. Thus we can assume $a_n \to 0$ and we have to prove the corresponding statement for the averages.

Take any illimited $n \in \mathbb{N}$. Using nonstandard induction on m, we see that $(a_1 + \ldots + a_m)/n \simeq 0$ for all standard $m \in \mathbb{N}$.

By Robinson's lemma, there is a nonstandard m with the same property. Then, all $|a_i|$ for $m < i \leq n$ are infinitesimal and so must be their maximum. *A fortiori*

$$|a_{m+1} + \ldots + a_n|/n \leq |a_{m+1} + \ldots + a_n|/(n-m) \simeq 0.$$

3.5.13 If x is limited with $x^* \neq 0$, *a fortiori* $x \neq 0$ and we can write for example

$$|x| \simeq |x^*| > |x^*|/2 \quad \text{whence } |x| > |x^*|/2 \text{ standard}$$

and finally $|1/x| < 2/|x^*|$, i.e. $1/x$ is limited. If x and $1/x$ are limited, say for example $|x|$ and $|1/x| \leq a$ standard, then

$$-\log a \leq \log|x| \leq \log a$$

and thus $\log |x|$ is limited. Finally, if $\log |x|$ is limited, so is $|x| = \exp \log |x|$.

3.5.14 Assume x appreciable and $x \simeq y$. Thus $x - y = \varepsilon \simeq 0$ and

$$1 - y/x = \varepsilon(1/x) \simeq 0 \text{ (since } 1/x \text{ is limited): } x \approx y.$$

Conversely, if x is limited and $x \approx y$,

$$1 - y/x = \varepsilon \simeq 0 \Rightarrow x - y = x\varepsilon \simeq 0 : x \simeq y.$$

Let us write the second part of the proof in the complex case. If $a_i \approx b_i$ for all i, then for each standard $\varepsilon > 0$, we have

$$|1 - a_i/b_i| < \varepsilon$$

and thus

$$\sum|b_i - a_i| < \varepsilon \sum|b_i|,$$

$$|\sum b_i - \sum a_i|/|\sum b_i| < \varepsilon.$$

This shows

$$\sum b_i \approx \sum a_i.$$

3.5.15 Let the galaxy $\mathscr{A}$ be defined by the function f (resp. $\mathscr{B}$ by g). Since they are disjoint, we have

$$f(x) \text{ limited} \Rightarrow g(x) \text{ illimited},$$
$$g(x) \text{ limited} \Rightarrow f(x) \text{ illimited}.$$

Define

$$A = \{x \in E : f(x) < g(x)\} \quad \text{and} \quad B = \{x \in E : f(x) > g(x)\}.$$

These subsets are obviously disjoint and $x \in \mathscr{A}$ means $f(x)$ limited, hence $g(x)$ illimited and $f(x) < g(x) : x \in A$. Similarly $\mathscr{B} \subset B$.

3.5.16 Simple exercise of style . . . .

3.5.17 When b is appreciable (Exercise 3.5.13), E^* is the ellipse of equation $x^2/a^2 + y^2/(b^*)^2 = 1$. When b is illimited, E^* is the union of the two horizontal lines $x = \pm a$. Finally, when b is infinitesimal E^* is the interval $[-a, a]$ of the real axis.

3.5.18 (a) Let x be a standard extremity of C, i.e. x is an end point of an interval of a C_m for a standard m. It is obvious that $x \in C_m$ for all $m' > m$ and in particular $x \in C_n$ (n illimited). This proves that $x \in C_n^*$. By transfer, the set of all extremities of C (this set is dense in C) is contained in the (standard) set C_n^*. Since this last set is closed, we conclude $C \subset C_n^*$.

(b) Conversely, if x is standard, the sequence $m \mapsto d(x, C_m)$ is standard too. But if x is a standard element of C_n^*, we have $d(x, C_n^*) \simeq 0$. This proves that $d(x, C_m) \to 0$ and $x \in \cap C_m = C$.

This set C is the **Cantor set**. It is the simplest nontrivial example of a **fractal**. The interested reader will also be able to construct the 'von Koch curve' with NSA as follows. The first approximation is given by installing an equilateral triangle with base equal to the middle third of the segment $[0, 1]$ and deleting this middle third as in the Cantor set construction. This first approximation consists thus of four connected segments (cf. next figure) of equal length $1/3$. Repeat the construction above each of them to obtain a curve consisting of $16 = 4^2$ segments of equal length $1/9$. The process is easily continued by induction, and the **von Koch curve** (already considered in 1904) is the limit, or the standard part of an illimited step.

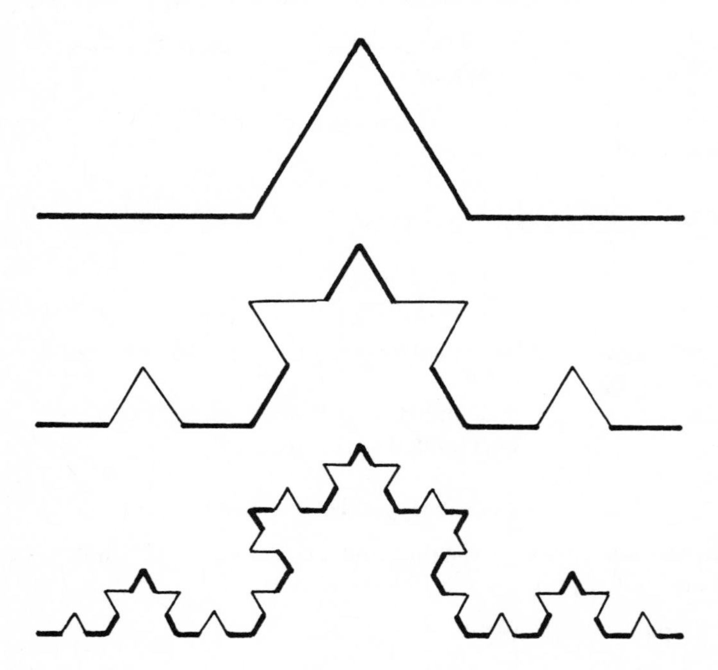

Fig. 3.3

3.5.19 Since A is standard in X (also standard), the set of sequences $(a_n) \subset A$ which converge to $x \in X$ is standard whenever x is standard. This set is not empty if $x \in \bar{A}$, hence contains some standard elements (Exercise 2.8.1).

CHAPTER 4

4.6.1 The function x^n, sin nx, e^x are continuous at all points x (and for all integers n).

(a) If x is standard and $|x| < 1$, the sequence $n \mapsto x_n$ is standard and tends to 0. Hence $x^n \simeq 0$ for illimited n (and x as above). This proves that $x \mapsto x^n$ (n illimited) is S-continuous at all limited points x with $|x^*| < 1$. But this function is not S-continuous at points x with $|x| \geq 1$. For example, $x = 1 - 1/n \simeq 1$ but $x^n = (1 - 1/n)^n \simeq 1/e$ is not infinitely close to $(x^*)^n = 1$. More generally, if h is infinitesimal and $y = x + h \simeq x$, we have

$$f(y) - f(x) = (x+h)^n - x^n = nx^{n-1}h + \ldots + h^n$$
$$\geq nhx^{n-1} \geq nh \quad \text{if } x \geq 1 \quad \text{and} \quad h > 0.$$

This difference can be appreciable ($h = 1/\sqrt{n}$ makes it illimited).

(b) At the origin, the function sin nx behaves like nx and it cannot be S-continuous (we leave the reader to find trigonometrical formulas permitting to prove this without having recourse to differentiability). The same is true for all integral multiples of π/n. Finally, let x be arbitrary: there is an integral multiple of π/n which is infinitely close to x. Consequently, the function $x \mapsto \sin nx$ (n illimited) is S-continuous nowhere.

(c) The exponential function e^x is not S-continuous at illimited points $x > 0$. Take indeed such a point x and consider $h \simeq 0$ with

$$y = x + \ln(1+h) \simeq x$$

(the ln function is continuous and S-continuous at the standard point 1). We then have

$$e^y = e^x \cdot e^{\ln(1+h)} = e^x \cdot (1+h),$$
$$e^y - e^x = he^x \quad \text{not always infinitesimal}$$

(for example, $h = e^{-x}$ makes this difference equal to 1).

The reader can still show that the function $f(x) = 1/x$ defined for $x \neq 0$ is not S-continuous at points $x \simeq 0$, $x \neq 0$.

4.6.2 Fix $a < b$ and prove $f(a) \leq f(b)$. Take the integer n sufficiently large to have $(b-a)/n \simeq 0$ (if a and b are standard, it is enough to take n illimited, but in general, use the Archimedean property of $\mathbb{R}$ to show that such an n exists). Introduce the subdivision points

$$x_k = a + k(b-a)/n \quad (k \geq 1).$$

Then $f(a) \leq f(x_k)$ can be proved by induction on k, ($f(s) \leq f(t)$ is a classical formula).

4.6.3 We can take a finite set F containing all standard real numbers and put $n = \text{Card}(F)$ (illimited integer). There exists a polynomial p of degree $< n$ having the required properties. Polynomial functions are continuous functions (still true . . .). Let us show that if a continuous function f satisfies

$$f(r) = 1 \quad \text{for all standard rationals } r,$$
$$f(s) = 0 \quad \text{for all standard irrationals } s,$$

then f is not S-continuous at limited points. Let us examine first the case of a standard rational r. The sequence $n \mapsto r + \sqrt{2}/n$ is standard, consists of irrational numbers and tends to r. The set of integers n such that $f(r + \sqrt{2}/n) = 0$ contains all standard n, hence contains some nonstandard n as well (principle 2.4.4). For such an illimited integer n,

$$f(r) = 1 \quad \text{but } f(r + \sqrt{2}/n) = 0 \quad \text{in spite of } r + \sqrt{2}/n \simeq r.$$

At a standard irrational s, f is not S-continuous either. This is shown for example by consideration of the sequence $n \mapsto [ns]/n$ which is standard if s is, and consists of rational numbers tending to s. Finally, if x is limited $x^* = s$ is standard and S-continuity at x is equivalent to S-continuity at s.

4.6.4 When I is compact, the function f is uniformly continuous on I, however, we cannot infer that $f(x) \simeq f(y)$ as soon as $x \simeq y$ (counterexamples are given by $f(x) = \sin nx$ where n is illimited, or by the functions of the preceding exercise). Thus the condition $|x - y| < 1/n$ is not sufficient. But it is enough to add the hypothesis that f is standard. In this case, uniform continuity indeed gives

$$x \simeq y \Rightarrow f(x) \simeq f(y).$$

For $k = [x/\varepsilon]$, $k\varepsilon \simeq x$ and $g(x) = f(k\varepsilon) \simeq f(x)$ as desired.

4.6.5 Take a finite part F containing all standard numbers. If $N = \text{Card } F$ (illimited), the elements of F can be parametrized by the interval $[0, N[$: say $F = \{a_n : 0 \leq n < N\}$. Define then

$$f(x) = \sum 1/2^{n+1} \quad \text{finite sum extended over the } n \text{ with } a_n < x.$$

This function 'jumps' at all $a_n \in F$ hence is discontinuous at all standard points. This function is positive, increasing and bounded by

$$\sum_{n \in \mathbb{N}} 1/2^{n+1} = 1/2 + 1/4 + \ldots = 1.$$

4.6.6 Since the sum of two infinitesimals is also infinitesimal, the sum of two S-continuous functions is S-continuous. The same is true for a finite sum having a limited number of terms (as is seen by nonstandard induction). Be more careful about products, since at points where one factor is illimited, infinitesimal jumps of the other are magnified (take $f(x) = n$ illimited constant and $g(x) = 0$ for $x \leq 0$ whereas $g(x) = 1/n$ for $x > 0$: these functions are S-continuous at the origin but their product is not S-continuous at the same point).

4.6.7 (a) For each $\varepsilon > 0$, choose $\delta_\varepsilon > 0$ with

$$|y-x| < \varepsilon \Rightarrow |f(y)-f(x)| < \delta_\varepsilon.$$

Take an arbitrary z and put $c = |f(z)-f(x)|$, $d = |z-x|$. Then

$$\begin{aligned} |y-z| < \varepsilon \Rightarrow |y-x| &\leq |y-z|+|z-x| < \varepsilon+d \\ &\Rightarrow |f(y)-f(x)| < \delta_{\varepsilon+d} \\ &\Rightarrow |f(y)-f(z)| < c+\delta_{\varepsilon+d}. \end{aligned}$$

This proves that f is suounitnoc at z as soon as it is at x. We shall simply say that f is suounitnoc at one point (hence at all points also).

(b) When $\delta > 0$ has the property of the definition of suounitnoc with respect to some $\varepsilon > 0$, it is obvious that the same δ will also work for smaller $\varepsilon' \leq \varepsilon$. The point is thus to consider big values for ε. The intuitive formulation of suounitnoc is thus easily seen to be

$$f \quad \text{bounded on all bounded intervals.}$$

For example, the function $f(x) = x$ is suounitnoc. Constant functions are also suounitnoc. When c is illimited, $f(x) = c$ is suounitnoc but $f(0) = c$ is not limited and the criterium (c) can only be applied to standard functions.

(c) Let us show that if f is standard

$$f \quad \text{suounitnoc} \Leftrightarrow f(x) \quad \text{limited for all } x \text{ limited.}$$

By (b), we see that f is suounitnoc precisely when

$$\text{for each interval } I,\ I \text{ bounded} \Rightarrow f(I) \text{ bounded.}$$

When f is standard, transfer can be applied and the preceding condition is equivalent to

$$\text{for each standard interval } I,\ I \text{ bounded} \Rightarrow f(I) \text{ bounded.}$$

But by definition,

$$a \text{ limited} \Leftrightarrow a \text{ belongs to a standard bounded interval.}$$

Taking successively $a = x$ and then $a = f(x)$, we obtain the desired equivalence.

CHAPTER 5

5.6.1 Take an illimited n and a limited x. Then $x/n \simeq 0$ and

$$\log(1+x/n) = x/n + \varepsilon x/n \quad \text{where } \varepsilon \simeq 0.$$

Hence

$$(1+x/n)^n = \mathrm{e}^{n \log(1+x/n)} = \mathrm{e}^{n(x/n+\varepsilon x/n)}$$
$$= \mathrm{e}^{x+\varepsilon x} = \mathrm{e}^x \, \mathrm{e}^{\varepsilon x} \simeq \mathrm{e}^x$$

since $\varepsilon x \simeq 0$ and $t \mapsto \mathrm{e}^t$ is S-continuous at $t=0$.

5.6.2 If f is S-differentiable at the point a, there is a standard m with

$$\frac{f(x)-f(a)}{x-a} = m+\varepsilon \quad \text{with } \varepsilon=\varepsilon(x)\simeq 0 \quad \text{for all } x\simeq a.$$

Hence

$$f(x)-f(a)=(m+\varepsilon)(x-a)\simeq 0 \quad \text{for } x\simeq a.$$

5.6.3 It is classical that f_ε is differentiable with derivative 0 at all non multiples of ε (this is true for *all* ε). At multiples of ε, f_ε is not continuous and *a fortiori* not differentiable.

CHAPTER 6

6.4.1 By definition

$$\int_0^1 e^x \, \mathrm{d}x = \mathrm{st} \sum_{0 \le jh < 1} e^{jh} h = \mathrm{st}(1+e^h+ \ldots + e^{(n+1)h}).$$

It is easy to sum this geometric progression with ratio h, finding

$$\mathrm{st}\, h\frac{e^{nh}-1}{e^h-1} = (e-1)\, \mathrm{st}\, h/(e^h-1) \quad (\text{since } nh=1).$$

But we have seen that the exponential is differentiable with derivative e^x. In particular, at the origin, the derivative is 1 and the desired standard part is 1. We have found

$$\int_0^1 e^x \, \mathrm{d}x = e-1.$$

The second integral can be computed in a similar fashion, observing that the sum

$$r+2r^2+3r^3+ \ldots + nr^n = r(1+2r+ \ldots + nr^{n-1})$$

can be computed by derivation of the sum of the geometric series

$$1+2r+3r^2+\ldots nr^{n-1} = (r+r^2+\ldots+r^n)'$$

is the derivative of

$$r(r^n-1)/(r-1).$$

6.4.2 Proceeding as above with $h \simeq \pi/n$ (n illimited), the proposed integral is the standard part of

$$\sum_{0 \le j < n} \log(1-2a \cos j\pi/n + a^2)\cdot \pi/n$$

$$= \frac{\pi}{n} \log \prod_{0 \le j \le n-1} (a-e^{ij\pi/n})(a-e^{-ij\pi/n})$$

$$= \frac{\pi}{n} \log \left[\frac{a-1}{a+1}(a^{2n}-1)\right].$$

(The product corresponds to $2n$th-roots of 1, $a=1$ appearing twice and $a=-1$ being absent.) We see then that the standard part is

$$\pi \log a^2 \quad \text{if } a>1, \quad 0 \text{ otherwise.}$$

Indeed,

$$\int_0^\pi \log(1-2a \cos x + a^2)\mathrm{d}x = \begin{cases} \pi \log a^2 & \text{if} |a|>1 \\ 0 & \text{if} |a|<1 \end{cases}.$$

6.4.3 Take $x \simeq 1$ and $x<1$ so that

$$\text{st} \log(1+x) = \log(1+\text{st}\, x) = \log 2$$

(the standard function log is continuous, hence also S-continuous at the standard point $x=1$). On the other hand, choose an illimited integer N so that

$$\text{st}\,(x - x^2/2 + \ldots) = \text{st} \sum_{n<N} (-1)^{n-1} x^n/n.$$

We have to prove that

$$\sum_{n<N} (-1)^{n-1} x^n/n \simeq \sum_{n<N} (-1)^{n-1}/n$$

since then we shall have

$$\text{st} \sum_1^\infty (-1)^{n-1} x^n/n = \text{st} \sum_{n<N} (-1)^{n-1}/n = \sum_1^\infty (-1)^{n-1}/n.$$

Let us estimate the difference

$$\sum_{n<N} (-1)^{n-1} x^n/n - \sum_{n<N} (-1)^{n-1}/n = -\sum_{n<N} (-1)^{n-1}(1-x^n)/n.$$

For this, let us write

$$M_n(x) = (1 + x + \ldots + x^{n-1})/n \quad \text{(average of powers of } x\text{)}$$

so that the above difference can be written

$$(1-x)\sum_{n<N}(-1)^{n-1}M_n(x).$$

The numbers $M_n(x)$ tend to zero monotonously so that their alternating sum is majorized by the first term 1. The conclusion results from the assumption $x-1\simeq 0$.

More generally, one can show

$$(*)\ \sum_{n<N}a_nx^n\simeq\sum_{n<N}a_n\quad\text{if}\quad x<1\quad\text{and}\quad x\simeq 1$$

as soon as

$$\left|\sum_{1\leq n<p}na_n\right|\leq\sigma\quad\text{limited (independent from } p).$$

Indeed, the difference is as above

$$(1-x)\sum_{n<N}na_nM_n(x).$$

Abel's transformation shows that $\sum_{n<N}na_nM_n(x)$ is limited.

When (*) is satisfied, we conclude

$$\sum_1^{\infty}a_n=\text{st}\sum_{1\leq n<N}a_n=\text{st}\sum_{1\leq n<N}a_nx^n$$

with

$$\sum_{1\leq n<N}a_nx^n\simeq f(x)=\sum_1^{\infty}a_nx^n$$

if this last series converges. In particular, when the sequence $n\mapsto a_n$ is standard (i.e. f standard)

$$\sum_1^{\infty}a_n=\text{st}\,f(x)=f(\text{st}\,x)=f(1).$$

CHAPTER 7

7.5.1 The part $A=[-n,n]$ is finite for each n, hence $m(A)=0$ but when n is illimited, A is not standard and $m(A)$ is not equal to the standard part of $m_n(A)$.

7.5.2 The first assertion is obvious. But the equality $m(\mathbb{N})=1/2$ is only valid for the special invariant means defined by the procedure used in the text. For example, one could define

$$m'_n(A)=(1/n)\text{ Card }(A\cap[0,n[)$$

and then, fixing an illimited n

$$m'(A) = \text{st } m'_n(A) \quad \text{for all standard } A.$$

The reader will prove that these m' are also invariant means on $\mathbb{Z}$ and that for example

$$m'(\mathbb{N}) = m'([k, \infty[) = 1,$$

$$m'(]-\infty, k]) = 0$$

for all $k \in \mathbb{Z}$. There are—of course—still other invariant means on $\mathbb{Z}$ . . .

7.5.3 The part $3A$ consists of the integers $3m$ where

$$3^{2i+1} < 3|m| \leq 3^{2i+2} \quad \text{for some integer } i \in \mathbb{N}.$$

Let $A_+ = A \cap \mathbb{N}$. It is obvious that the three parts

$$3A_+, \; 3A_+ - 1 \quad \text{and} \quad 3A_+ - 2$$

constitute a partition of $A'_+ = A' \cap \mathbb{N}$. Hence $3m(3A_+) = m(A'_+)$ and similarly for the negative part of A. This proves that

$$m(3A) = m(A')/3.$$

In the case $m = \text{st } m_n$ with $n = 3^{2k}$, we have

$$m(A) = 1/4, \quad \text{hence } m(A') = 3/4 \quad \text{and} \quad m(3A) = 1/4 \; (= m(A)!).$$

If, on the contrary, n is an odd power of 3, we have

$$m(A) = 3/4, \quad \text{hence } m(A') = 1/4 \quad \text{and} \quad m(3A) = 1/12 \; (\neq m(A)/3!).$$

7.5.4 First, m is not countably additive on parts of $\mathbb{Z}^2$ ($\mathbb{Z}^2$ is itself a countable union of points, each of which has zero invariant mean). Second, the inequality $m(dA) = m(A)/d^2$ is not valid for all parts $A \subset \mathbb{Z}^2$. . . .

However, the conclusion is correct as the following argument (due to Cesaro) shows. Let us work with the mean on $\mathbb{N}^2$ defined as in the preceding exercise (and defining an invariant mean on $\mathbb{Z}^2$)

Q is the square $[1, n]^2$ where n is a fixed illimited integer,

$$m_n(A) = n^{-2} \text{ Card } (A \cap Q),$$

$$m(A) = \text{st } m_n(A) \quad \text{for all standard parts } A \subset \mathbb{N}^2.$$

(This choice will have little importance—as we shall see—but is convenient since summations can then be extended over positive values for the integers k: this will be tacitly assumed.)

Let us compute the density of $Q_1 = Q \cap A_1$ by estimating the number of couples of relatively prime integers in Q as follows. To obtain Card Q_1, it is necessary to subtract from Card Q the number of couples multiples of 2, 3, . . . (only taking into account primes $p \leq n$). As there are $[n/p]^2$ couples which are multiples of p in Q, a first approximation for the desired number is

$$\text{Card } Q_1 : n^2 - \sum_{p \leq n} [n/p]^2.$$

Effecting this difference, we have subtracted twice the couples which are a multiple of 6, . . . more precisely, we have subtracted twice the couples having a gcd multiple of a product $pq \leq n$ of two distinct primes p and q. We have to add a corrective term for the number of these couples. This term is

$$[n/pq]^2 \quad \text{for primes } p < q \quad \text{with } pq \leq n.$$

Finally, we obtain a correct formula by writing the alternating sum

$$\text{Card } Q_1 = n^2 - \sum_{p \leq n} [n/p]^2 + \sum_{p < q, pq \leq n} [n/pq]^2 - \dots .$$

The density of Q_1 in Q is

$$n^{-2} \text{ Card } Q_1 = 1 - \sum n^{-2}[n/p]^2 + \sum n^{-2}[n/pq]^2 - \dots \tag{1}$$

(with the same restrictions on primes in these sums). We shall compute this density by comparing it to the alternating sum

$$1 - \sum 1/p^2 + \sum 1/p^2q^2 - \dots \text{ (same } p, q, \dots \text{ as in (1))}. \tag{2}$$

This second sum is indeed easy to estimate. Forgetting the restricting $pq \leq n$, $pqr \leq n, \dots$ and keeping only in mind that $p \leq n$ in the first sum, $p < q \leq n$ in the second, . . . we add terms (up to sign) taken from the tail of the standard convergent series Σk^{-2}:

$$\sum_{k > n} 1/k^2 \simeq 0 \quad \text{(since } n \text{ is illimited).}$$

This shows that (2) is infinitely close to

$$1 - \sum_{p \leq n} p^{-2} + \sum_{p < q \leq n} p^{-2}q^{-2} - \dots = \prod_{p \leq n} (1 - p^{-2})$$

and also infinitely close to the convergent infinite product

$$\prod_{p \text{ prime}} (1 - p^{-2}) = 1/\zeta(2) = 6/\pi^2. \tag{3}$$

It only remains to show that the difference between (1) and (2) is infinitesimal: this will give $m(A_1) = 6/\pi^2$ independently from the choice of the illimited integer n chosen to define m. For this we have to estimate the differences

$$1/k^2 - [n/k]^2/n^2$$

and take their sum for $k \leq n$. By definition of the integral part, we have

$$n = [n/k]k + r \quad \text{with } 0 \leq r = r_k < n.$$

Hence

$$1/k = [n/k]/n + r/nk,$$
$$1/k^2 - 2r/(nk^2) + r^2/(n^2k^2) = [n/k]^2/n^2,$$

and thus

$$0 \leq 1/k^2 - [n/k]^2/n^2 \leq 2r/(nk^2) - \ldots < 2r/(nk^2) < 2/nk.$$

The sum of these differences is majorized by

$$\sum_{k \leq n} 2/nk = 2n^{-1} \sum_{k \leq n} 1/k. \tag{4}$$

But it is clear that the averages $M_n = n^{-1} \sum_{1 \leq k \leq n} 1/k$ of the sequence $1/k$ (monotonously decreasing to 0) are infinitesimal for n illimited (cf. Exercise 3.5.12).

CHAPTER 8

8.5.1 (a) Here is an example of a continuous Dirac function which does not satisfy $D(x) \simeq 0$ for all noninfinitesimal $x \in \mathbb{R}$:

$$D(x) = \begin{cases} \text{peak of surface } 1-\varepsilon, \text{ height } 1/\varepsilon, \text{ supported in } |x| \leq \varepsilon \\ \text{peak of surface } \varepsilon, \text{ height 1, supported in } |x-1| \leq \varepsilon \\ \text{vanishes otherwise.} \end{cases}$$

(Make a picture!)

(b) Look at Proof (8.1.3) and insert absolute values where needed!

8.5.2 Let $E = [0, 1] \supset A = \{0, 1/n, 2/n, \ldots, n/n = 1\}$ where n is an illimited integer. Then A is finite hence closed in E. But each $x \in E$ lies between two consecutive multiples of $1/n$

$$k = [nx] \Rightarrow k \leq nx < k+1 \Rightarrow$$
$$k/n \leq x < (k+1)/n \Rightarrow |x - k/n| \leq 1/n \simeq 0.$$

Thus, (ii) is true but (i) fails. Observe that (i)$\Rightarrow$(ii) in general: if A is dense in E and $x \in E$, for each illimited integer $n \in \mathbb{N}$ there exists $a \in A$ with $d(x, a) \leq 1/n \simeq 0$.

When A is standard, (ii)$\Rightarrow$(i) by transfer.

One further warning, my son: the use of books is endless, and much study is wearisome.
(Ecclesiastes)

BIBLIOGRAPHY

BOOKS

[B] E. Borel, *Les Nombres Inaccessibles*, Gauthier–Villars, Paris (1952).

[C] P. J. Cohen, *Set Theory and the Continuum Hypothesis*, W. A. Benjamin Inc. New York (1966).

[D] J. Dieudonné, *Foundations of Modern Analysis*, Academic Press, New York (1960).

[Di] F. Dienes, *Cours d'analyse non standard*, Publ. Univ. d'Oran (1983), (épuisé).

[E] L. Euler, *Opera Omnia, Introductio in Analysin Infinitorum*, Tomi Primi, Lausanne (1748).

[H] P. R. Halmos, *Naïve Set Theory*, van Nostrand Co. (1960).

[HL] A. E. Hurd, and P. A. Loeb, *An Introduction to Non Standard Real Analysis*, Academic Press (1985).

[K] J. H. Keisler, *Elementary Calculus*, Prindle, Weber & Schmidt, Boston (1976).

[K′] J. H. Keisler, *Foundations of Infinitesimal Calculus*, Prindle, Weber & Schmidt, Boston (1976).

[L] R. Lutz, and M. Goze, *Non Standard Analysis*, Springer Lect. Notes in Mathematics, #881 (1981).

[La] D. Laugwitz, *Zahlen und Kontinuum* (*Eine Einfuhrung in die Infinitesimalmathematik*), B. I.-Wissenschaftsverlag Mannheim/Wien/ Zürich (1986).

[N] E. Nelson, *Radically Elementary Probability Theory* (in preparation).

[R] A. Robinson, *Non Standard Analysis*, North-Holland Publ. Co. (1966).

ARTICLES

[1] Aronszajn, N., and Smith, K. T., Invariant subspaces of completely continuous operators, *Ann. Math.*, **60** (1964), 345–350.

[2] Bernstein, A. R., and Robinson, A., Solution of an invariant subspace problem of K. T. Smith and P. R. Halmos, *Pacific Journal of Math.*, **16** (3) (1966), 421–431.

[3] Cartier, P., Perturbations singulières des équations différentielles ordinaires et analyse non standard, *Sém. Bourbaki*, Nov. 1981, Astérisque 92–93 (1982), 21–44.

[4] Halmos, P. R., Invariant subspaces of polynomially compact operators, *Pacific Journal of Math.*, **16**(3) (1966), 433–437.

[5] Hilden, H. M., Proof of Lomonosov's theorem, *Math. Intelligencer*, **4**(1) (1982), 34.

[6] Hurd, A. E., ed., Nonstandard analysis, recent developments, Springer Lect. Notes in Mathematics, Nb. 983 (1983).

[7] Lakatos, I., The Significance of Non Standard Analysis for the History and Philosophy of Mathematics, *Math. Intelligencer*, **1**(3) (1978), 151–161.

[8] Laugwitz, D., ed., Nichtstandard analysis, *Math. Unterricht*, **29**(4) (1983).

[9] Lomonosov, V. J., Invariant subspaces for operators commuting with compact operators. *Funct. Anal. Appl.*, **7** (1973), 213–215.

[10] Nelson, E., Internal set theory, a new approach to NSA, *Bull. Amer. Math. Soc.*, **83** (1977), 1165–1198.

[11] Richter, M., Ideale Punkte, Monaden und Nichtstandard Methoden, Braunschweig, Wiesbaden: Vieweg (1982).

[12] Robert, A., Une approache naïve de l'ANS, *Dialectica* 38 *fasc.* **4** (1984), 287–296.

[13] Robert, A., L'Analyse Non Standard, L'Encyclopédie Philosophique, Vol. 1, Presses Univ. France, Paris (1987).

[14] van den Berg, I., Nonstandard asymptotic analysis, Springer Lect. Notes in Mathematics, Nb. 1249 (1987).

INDEX

BASIC PRINCIPLES OF NSA

1. In every *infinite* set, there are some nonstandard elements (1.4.4).
   Equivalently:
   A set which contains only standard elements is *finite*.

2. In any set, there is a *finite part* which contains all standard elements of E.

3. Let E be a *standard infinite* set. Then:
   (a) each subset A of E which contains all standard elements of E also contains some nonstandard elements (2.4.4).
   (b) each subset A of E which contains all nonstandard elements of E also contains some standard elements.

4. Let P be any property (classical or not) defined for all natural integers. Assume

   - $P(0)$ is true,
   - $\forall^s n \quad (P(n) \Rightarrow P(n+1))$.

   Then $P(n)$ is true for all standard $n \in \mathbb{N}$ (2.8.4).

5. Robinson's Lemma. Let $(a_n)_{n\in\mathbb{N}}$ be a sequence of real (or complex) numbers. If $a_n \simeq 0$ for every standard $n \in \mathbb{N}$, then there exists an illimited integer $\nu \in \mathbb{N}$ with $a_n \simeq 0$ for all $n \leq \nu$.

6. Let E and F be two standard sets. If a construction (classical or not) furnishes a standard element $f(x) \in F$ for each standard element $x \in E$, then there is a unique standard map $f: E \rightarrow F$ taking the prescribed values for standard $x \in E$. In particular, if a standard element $a_n \in F$ is constructed for each standard $n \in \mathbb{N}$, then there is a unique standard sequence $(a_n)_{n\in\mathbb{N}}$ in F extending the preceding definition.

IST AXIOMS FOR NSA

(I) IDEALIZATION

Let $R = R(x, y)$ be a *classical* relation.

In order to be able to find an x with $R(x, y)$ for all standard y,

a necessary and sufficient condition is

for each standard finite part F, it is possible to find an $x = x_F$ such that $R(x, y)$ holds for all $y \in F$.

(S) STANDARDIZATION

Let E be a *standard* set and P any property.

There exists a (unique) standard subset A of E, denoted by ${}^S\{x \in E : P(x)\}$ having the following property: the standard elements of A are precisely the standard elements of E satisfying P.

(T) TRANSFER

Let F be a *classical* formula in which all parameters $A, B, \ldots L$ have some fixed *standard* values.

$F(x, A, B, \ldots L)$ is true for all x as soon as it is true for all standard x.

Equivalently

If there exists an x such that $F(x, A, B, \ldots L)$ is true, there also exists a standard x such that $F(x, A, B, \ldots L)$ is true.

In particular

If there is a unique x such that $F(x, A, B, \ldots L)$ is true, then this x must be standard.